Lecture Notes in Computer Sc

T0237812

Commenced Publication in 1973
Founding and Former Series Editors:
Gerhard Goos, Juris Hartmanis, and Jan van Leeuwen

Hendrik Jan Hoogeboom   Gheorghe Păun
Grzegorz Rozenberg   Arto Salomaa (Eds.)

Membrane Computing

7th International Workshop, WMC 2006
Leiden, The Netherlands, July 17-21, 2006
Revised, Selected, and Invited Papers

 Springer

Volume Editors

Hendrik Jan Hoogeboom
Leiden Center of Advanced Computer Science (LIACS)
Leiden University
Niels Bohrweg 1, 2333 CA Leiden, The Netherlands
E-mail: hoogeboom@liacs.nl

Gheorghe Păun
Institute of Mathematics of the Romanian Academy
PO Box 1-764, 014700 Bucureşti, Romania, and
Research Group on Natural Computing
Department of Computer Science and AI
Seville University, 41012 Seville, Spain
E-mail: george.paun@imar.ro, gpaun@us.es

Grzegorz Rozenberg
Leiden Center of Advanced Computer Science (LIACS)
Leiden University
Niels Bohrweg 1, 2333 CA Leiden, The Netherlands
E-mail: rozenberg@liacs.nl

Arto Salomaa
Turku Centre for Computer Science (TUCS)
Leminkäisenkatu 14, 20520 Turku, Finland
E-mail: asalomaa@cs.utu.fi

Library of Congress Control Number: 2006939014

CR Subject Classification (1998): F.1, F.4, I.6, J.3

LNCS Sublibrary: SL 1 – Theoretical Computer Science and General Issues

ISSN   0302-9743
ISBN-10   3-540-69088-3 Springer Berlin Heidelberg New York
ISBN-13   978-3-540-69088-7 Springer Berlin Heidelberg New York

Springer is a part of Springer Science+Business Media

springer.com

© Springer-Verlag Berlin Heidelberg 2006
Printed in Germany

Typesetting: Camera-ready by author, data conversion by Scientific Publishing Services, Chennai, India
Printed on acid-free paper   SPIN: 11963516   06/3142   5 4 3 2 1 0

Preface

This volume of the Springer *Lecture Notes in Computer Science* series contains the contributions presented at the International Workshop on Distributed, High-Performance and Grid Computing in Computational Biology 2006 (GCCB 2006) held in Eilat, January 21, 2007 in conjunction with the fifth European Conference on Computational Biology (ECCB 2006).

Modern computational biology and bioinformatics are characterized by large and complex-structured data and by applications requiring considerable computing resources, such as processing units, storage elements and software programs. In addition, these disciplines are intrinsically geographically distributed in terms of their instruments, communities and computing resources. Tackling the computational challenges in computational biology and bioinformatics increasingly requires high-end and distributed computing infrastructures, systems and tools. The main objective of this workshop is to bring together researchers and practitioners from these areas to discuss ideas and experiences in developing and applying distributed, high-performance and grid computing technology to problems in computational biology and bioinformatics.

The challenges in distributed, high-performance and grid computing in biology and biotechnology are inherently more complicated than those in such domains as physics, engineering and conventional business areas. Some of the added complexities arise from the:

- Conceptual complexity of biological knowledge and the methodologies used in biology and biotechnology
- Need to understand biological systems and processes at a detailed mechanistic, systemic and quantitative level across several levels of organization (ranging from molecules to cells, populations, and the environment)
- Growing availability of high-throughput data from genomics, transcriptomics, proteomics, metabolomics and other high-throughput methods
- Widespread use of image data in biological research and development (microscopy, NMR, MRI, PET, X-ray, CT, etc.)
- Increasing number of investigations studying the properties and dynamic behavior of biological systems and processes using computational techniques (molecular dynamics, QSAR/QSPR, simulation of gene-regulatory, signaling and metabolic networks, protein folding/unfolding, etc)
- Requirement to combine data, information and compute services (e.g., sequence alignments) residing on systems that are distributed around the world
- Variety of different technologies, instruments, infrastructures and systems used in life science R&D
- Huge variety of information formats and frequent addition of new formats arising from new experimental protocols, instruments and phenomena to be studied

- Large and growing number of investigated biological and biomedical phenomena
- Fact that life science R&D is based heavily on the use of distributed and globally accessible computing resources (databases, knowledge bases, model bases, instruments, text repositories, compute-intensive services)

The GCCB workshop brought together computational biologists, bioinformaticians and life scientists who have researched and applied distributed, high-performance and grid computing technologies in the context of computational biology and bioinformatics. The workshop discussed innovative work in progress and important new directions. By sharing the insights, discussing ongoing work and the results that have been achieved, we hope the workshop participants conveyed a comprehensive view of the state of the art in this area and identified emerging and future research issues. We believe that the GCCB workshop made a valuable contribution in encouraging and shaping future work in the field of distributed, high-performance and grid computing in computational biology.

Acknowledgements

GCCB 2006 was sponsored and organized by the University of Ulster, Coleraine, UK; the Technion - Israel Institute of Technology, Haifa, Israel; the University of Amsterdam, Amsterdam, The Netherlands; the Dresden University of Technology, Dresden, Germany; and the DataMiningGrid Consortium (EC grant IST-2004-004475, www.DataMiningGrid.org). We are indebted to the ECCB 2006 Conference organizers Hershel Safer, Weizmann Institute of Science, and Haim Wolfson, Tel Aviv University. Finally, we would like to extend our gratitude to the members of the GCCB 2006 International Program Committee and to those who helped in the review process. And last but not least we want to thank all authors and workshop participants for their contributions and the valuable discussions.

January 2007   Werner Dubitzky
                                                          Assaf Schuster
                                                          Peter Sloot
                                                          Michael Schroeder
                                                          Mathilde Romberg

Organization

Program Chairs

Werner Dubitzky, University of Ulster, Coleraine, UK
Assaf Schuster, Technion - Israel Institute of Technology, Haifa, Israel
Peter M.A. Sloot, University of Amsterdam, Faculty of Sciences, Amsterdam, The Netherlands
Michael Schroeder, Dresden University of Technology, Biotechnological Centre, Dresden, Germany
Mathilde Romberg, University of Ulster, Coleraine, UK

Program Committee

David A. Bader, Georgia Tech, College of Computing, Atlanta, Georgia, USA
Eric Bremer, Children's Memorial Hospital, Northwestern University, Chicago, USA
Rui M. Brito, Universidade de Coimbra, Coimbra, Portugal
Marian Bubak, AGH - University of Science and Technology, Krakow, Poland
Kevin Burrage, The University of Queensland, Australia
Gustavo Deco, Universitat Pompeu Fabra, Barcelona, Spain
Frank Dehne, University of Ottawa, Ottawa, Canada
Guiseppe di Fatta, University of Konstanz, Konstanz, Germany
Werner Dubitzky, University of Ulster, Coleraine, UK
Jordi Vill i Freixa, Universitat Pompeu Fabra, Barcelona, Spain
David Gilbert, University of Glasgow, Glasgow, UK
Carol Goble, University of Manchester, Manchester, UK
Danilo Gonzalez, Universidad de Talca, Talca, Chile
Ulrich Hansmann, Forschungszentrum Juelich, NIC, Juelich, Germany
Des Higgins, University College Dublin, Conway Institute, Dublin, Ireland
Alfons Hoekstra, University of Amsterdam, Amsterdam, The Netherlands
Martin Hoffmann, Fraunhofer Institute for Algorithms and Scientific Computing SCAI, Sankt Augustin, Germany
Rod Hose, University of Sheffield, Sheffield, UK
Chun-Hsi (Vincent) Huang, University of Connecticut, Storrs, USA
Akihiko Konagaya, Riken Genomic Sciences Center, Yokohama City, Japan
Miron Livny, University of Wisconsin at Madison, Wisconsin, USA
Uko Maran, University of Tartu, Tartu, Estonia
Hartmut Mix, Dresden University of Technology, Dresden, Germany
Ron Perrot, Queens University, Belfast, UK
Mark Ragan, The University of Queensland, Australia

Table of Contents

Session 2a. "Data Management"

Session 2b. "Collaborative Environments"

Combining a High-Throughput Bioinformatics Grid and Bioinformatics Web Services

Chunyan Wang[1], Paul M.K. Gordon[1], Andrei L. Turinsky[1],
Jason Burgess[2], Terry Dalton[2], and Christoph W. Sensen[1]

[1] Sun Center of Excellence for Visual Genomics, University of Calgary,
HS 1150, 3330 Hospital Dr. NW, Calgary, Alberta, T2N 4N1, Canada
[2] National Research Council Institute for Marine Biosciences,
1411 Oxford Street, Halifax, Nova Scotia, B3H 3Z1, Canada
{cwan, gordonp, aturinsk}@ucalgary.ca,
{jason.burgess, terry.dalton}@nrc-cnrc.gc.ca, csensen@ucalgary.ca

Abstract. We have created a high-throughput grid for biological sequence analysis, which is freely accessible via bioinformatics Web services. The system allows the execution of computationally intensive sequence alignment algorithms, such as Smith-Waterman or hidden Markov model searches, with speedups up to three orders of magnitude over single-CPU installations. Users around the world can now process highly sensitive sequence alignments with a turnaround time similar to that of BLAST tools. The grid combines high-throughput accelerators at two bioinformatics facilities in different geographical locations. The tools include TimeLogic DeCypher boards, a Paracel GeneMatcher2 accelerator, and Paracel BlastMachines. The Sun N1 Grid Engine software performs distributed resource management. Clients communicate with the grid through existing open BioMOBY Web services infrastructure. We also illustrate bioinformatics grid strategies for distributed load balancing, and report several nontrivial technical solutions that may serve as templates for adaptation by other bioinformatics groups.

Keywords: Bioinformatics, sequence alignment, grid, Web services, BioMOBY, Sun N1 Grid Engine, Smith-Waterman, hidden Markov model.

1 Introduction

Analysis and functional annotation of genomic and proteomic sequences require fast and sensitive methods that are able to capture complex patterns within massive sequence collections in a reasonable time. The algorithms used to search for and rank sequence alignment regions, fall mostly into two classes: those based on dynamic programming, such as the Smith-Waterman algorithm [1] or the various BLAST methods [2]; and those based on probabilistic networks, such as hidden Markov model (HMM) searches [3]. The canonical Smith-Waterman or HMM alignments typically require massive computational power. For example, by incorporating hardware accelerators described in this paper into the MAGPIE genome annotation tool [4], more than 50 000 expressed sequence tags (EST), an

W. Dubitzky et al. (Eds.): GCCB 2006, LNBI 4360, pp. 1–10, 2007.

equivalent of six to eight full microbial genomes, can be annotated in one day. Without the accelerators, the annotation of one genome would require two to three weeks (unpublished data).

Most research teams cannot afford the local implementation of specialized high-throughput hardware, hence far less sensitive BLAST searches remain the most used type of sequence similarity searches today. Further hindrance to the research process is the disparity between data input specifications of many generic bioinformatics software tools [5, 6]. A typical bioinformatics analysis session likely involves switching between multiple tools, transferring data manually, and performing data format conversion to ensure compatibility.

To help overcome these challenges, we have created a high-throughput system for the alignment of biological sequences that combines the grid paradigm [7, 8] with the Web services paradigm [9]. There is a growing interest in distributed bioinformatics solutions, especially those that are Web services-oriented or grid-oriented [10, 11, 12]. Outside of bioinformatics, the grid and Web services-based computation methods are rapidly undergoing a standardization process [13]. Our system unites several hardware and software accelerators into a grid and provides universal access to it through an existing bioinformatics Web services interface, called BioMOBY [14]. The BioMOBY project endeavors to establish a *de facto* bioinformatics data interchange standard for the web, using open XML-based communication standards and protocols for service allocation and queries.

We provide access to several highly specialized sequence analysis hardware accelerators and software tools, while allowing dynamic load-balancing across geographically distributed computational nodes at the same time. Furthermore, the inner structure of our grid is entirely transparent to any BioMOBY user, and can be incorporated seamlessly into other BioMOBY data analysis pipelines via http://biomoby.org.

2   Materials and Methods

2.1 Architecture

Our architecture comprises three layers: the high-throughput accelerators, the grid middleware for job allocation, and the BioMOBY Web services interface (Fig. 1). At the bottom level, highly sensitive sequence alignments are performed using the resources of two bioinformatics facilities, which are three time zones apart: the Sun Center of Excellence for Visual Genomics at the University of Calgary; and the Institute for Marine Biosciences of the National Research Council Canada, in Halifax, Canada. The resources at the Calgary site include:

- Four FPGA-based (field programmable gate array) TimeLogic[TM] DeCypher® bioinformatics accelerator boards, mounted in a SunFire V880 server;
- A Paracel BlastMachine based on a Linux cluster with 12 dual-processor nodes (hereafter referred to as the 24-processor BlastMachine);

- An ASIC-based (application-specific integrated circuits) Paracel GeneMatcher2 bioinformatics accelerator with 27 648 specialized CPUs on nine boards; by design, jobs for the GeneMatcher2 are submitted through the Linux cluster;
- A Sun Fire E6900 UNIX server used for coordination tasks.

The resources at the Halifax side include:

- A Paracel BlastMachine based on a 12-node Linux cluster (hereafter referred to as the 12-processor BlastMachine);
- A dedicated Sun Enterprise 220R UNIX server used for the handling of command-line grid job submissions.

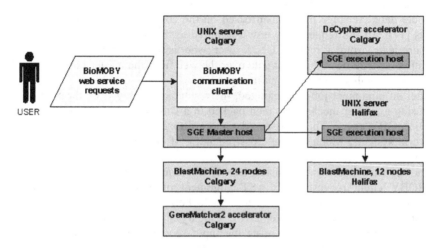

Fig. 1. The dataflow of the job submission to the BioMOBY-enabled grid test bed. The layered architecture comprises the computational nodes (*grey*), the grid middleware (*dark grey*) and the BioMOBY Web services components (*light grey*).

At the middle layer, the resources are united into a common wide area grid using Sun N1 Grid Engine 6 (SGE) middleware solution [15]. The SGE is installed on all machines as either a master host (main UNIX server in Calgary), which can submit jobs to other hosts, or an execution host (the remaining machines), which can only execute jobs locally. The BioMOBY-enabled SGE supports job execution over wide area networks, has a customizable policy module for formulating a flexible job submission policy, and allows easy out-of-the-box installation. The SGE defines internal queues for job scheduling, maintained at the master hosts, which may represent algorithm types, such as HMM searches, or individual execution hosts, or narrower job types, such as HMM Frame EST-to-HMM searches on the DeCypher system.

At the top layer, the BioMOBY Web services system provides a mechanism to discover the data processing resources and submit requests. Because all job submissions to the grid are handled by the main UNIX server in Calgary, the BioMOBY communication libraries only need to be installed on that host to create the

Web services endpoint. Using this endpoint, we registered the new BioMOBY web services, which are named for three types of sequence alignment analyses: runBlastVsPublicDBs, runSmithWatermanVsPublicDBs, and search_Pfam (for HMM searches). The registration process is described in [16]. Further information is available on the project website (http://moby.ucalgary.ca).

2.2 Data Flow Implementation

We had to overcome several unusual technical challenges in the context of implementing a geographically distributed grid test bed. First, traditional grids require users to log into their grid accounts to ensure remote authentication and secure data transfer. In contrast, BioMOBY analyses require no user accounts or authentication. Our solution was to create a specialized UNIX account on every SGE host and set up a remote data exchange mechanism based on the common gateway interface (CGI) scripts [17]. All job requests are now submitted to the SGE from this account, as if they originate from a regular UNIX user who is logged into the system. SGE execution hosts retrieve the input data over the web from the master host's CGI using the free Wget utility [18] and a unique input ID.

After an alignment job is completed, the results are retrieved by the SGE master host using a similar mechanism, then packed in a BioMOBY format and returned to the BioMOBY client. This CGI-based remote data exchange is in stark contrast to most grid systems, which require cross-mounting of disk drives between all hosts of the grid.

Another challenge was to translate BioMOBY requests into SGE submissions. If a BioMOBY client requests a BLAST alignment of a single sequence against a nucleotide sequence database, a custom-made Perl program parses the BioMOBY request at the main UNIX server at the Calgary site, which hosts the BioMOBY service endpoint. The program posts the input sequence as a separate file on the master host's CGI with a unique input identifier, and initializes a sequence-type parameter (nucleotide or amino acid). Subsequently it submits a pre-made SGE script all_blast.sh to the grid using the SGE submission command qsub, as follows:

```
qsub -hard -i numjobs=4 all_blast.sh $input_id $type
```

In this example, a simple threshold-based job scheduling is implemented. A SGE load sensor script goes through the list of hosts sorted in the order of their optimal BLAST performance. If it finds that the host has no more than four jobs in its SGE queue (numjobs=4), the new job of executing all_blast.sh is added to the host's queue. Otherwise the next best host from the list is examined.

The third challenge was to enable a generic SGE script to invoke job submissions with different syntax for different hosts. Continuing the BLAST example, the DeCypher and BlastMachine expect drastically different BLAST commands, whereas qsub invokes the same generic script all_blast.sh on either host. Our solution was to place on each host a local script blast.sh with BLAST commands specific to the host. A single all_blast.sh script can now trigger any of the local versions of blast.sh using the same command line, and pass on the required parameters.

3 Results and Discussion

Our objective was to explore possible grid load balancing policies that take into account the relative performances of the individual hosts. Even for single-sequence submissions, the grid considerably improves the performance of the alignment searches. For example, using the grid scheduling policy described in the previous section, we tested three types of alignment searches against public databases for a 599-amino-acid-long *Arabidopsis thaliana* calcium-dependent protein kinase AT3g19100/MVI11_1 (TrEMBL accession number Q9LJL9). The speedup was especially large for the computationally intensive Smith-Waterman searches, which took on average 5 100 seconds on a single CPU of the Sun Fire E6900 server from the Calgary site, but only 86 seconds on the grid – about 60 times faster (Fig. 2).

The benefits of parallel processing hardware become even more apparent when multiple input sequences are analyzed. This powerful option is available, for example, through the jMOBY Java-based toolkit for BioMOBY job submissions [19], plug-in modules for the Taverna workbench [20] as a front end to the BioMOBY system, or through other sequence analysis tools that support the BioMOBY Web services interface, such as [21]. At the time of this writing, the development of BioMOBY clients that access BioMOBY services or have an embedded BioMOBY functionality, is growing rapidly (for an up-to-date list see [22]). The features for the automated web-based and/or GUI-based submission of multiple sequences through BioMOBY clients, although a relatively new functionality, are also being constantly expanded.

For testing, we searched the NCBI Protein repository [23] for the keyword "alcohol" and retrieved the first N=100 sequences. This set was replicated to create the derivative sets of N=300, 1 000, 3 000, 10 000, 30 000, and 100 000 sequences, so that the values of log(N) cover the logarithmic scale almost uniformly between 2 and 5 with a step of approximately 0.5.

Alignments using a single CPU of the Sun Fire E6900 server in Calgary were tested on N=100 sequences, with the average execution time being 10 574 seconds for a BLAST, 3 961 seconds for an HMM search, and over 4 days(!) for a Smith-Waterman search. As Figure 2 shows, this constitutes a dramatic difference of two to three orders of magnitude compared to the grid performance on the same set of 100 sequences (listed in Table 1 for N=100). Tests for larger N values on a single CPU were not attempted due to the clear superiority of the grid and the specialized tools.

For each search type, the computational nodes exhibited near-linear performance with respect to the input size. However, we did not attempt linear regression because the data for smaller N are clearly outliers due to initial overhead and would cause linear least-squares fitting to underestimate consistently the "true" throughputs.

For BLAST searches, the throughputs were 4.25 seq/sec for the DeCypher, 1.13 seq/sec for the 24-processor BlastMachine, and 0.90 seq/sec for the 12-processor BlastMachine, measured at N=30 000, where deviations due to initial overhead were deemed minimal. We therefore split the input data among the three accelerators in proportion to their observed computational power: 68% for the DeCypher, 18% for the Calgary BlastMachine, and 14% for the Halifax BlastMachine. Generally, the load balancing should use the best available dynamic estimate for the throughput of each

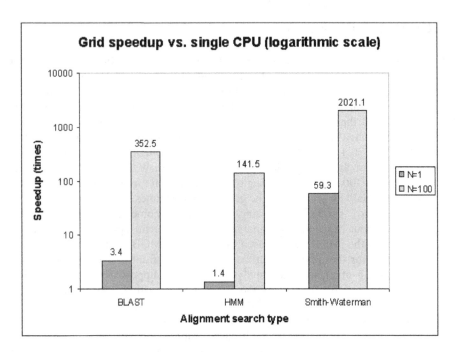

Fig. 2. The grid speedup compared to a single CPU on a Sun Fire E6900 server. Results reported in logarithmic scale for different numbers of input sequences and different alignment types (average of 6 trials).

computational node, taking into account such factors as the specific job type and search parameters. A proportional grid load balancing policy was defined:

Split large BLAST jobs among the DeCypher, Calgary BlastMachine, and Halifax BlastMachine in proportion to the (estimated) throughput of each accelerator.

For HMM searches, the DeCypher execution time ranged from 23 seconds for $N=10^2$ to 2 hours 43 minutes for $N=10^5$. In contrast, the execution time on the GeneMatcher2, which is a massively parallel supercomputer, stayed within the range of between 6 and 10 minutes. Using the approximate crossover point at $N=5\ 000$ between the two performance graphs, the grid load balancing policy was:

Submit the entire HMM job to the DeCypher if the input is less than 5 000 sequences, otherwise submit the entire job to the GeneMatcher2.

For Smith-Waterman searches, the GeneMatcher2 throughput was approximately 1.09 seq/sec for $N=30\ 000$. In contrast, the DeCypher throughput was only 0.07 seq/sec for $N=3\ 000$, with the execution time of nearly 12 hours. Due to the clear GeneMatcher2 superiority, DeCypher tests for larger N were not attempted. A trivial grid submission policy was defined:

For Smith-Waterman search requests, use the GeneMatcher2. Use the DeCypher only if the GeneMatcher2 is unavailable.

Table 1. Performance (in seconds) of different hosts compared to the grid load balancing policy. Hosts tested are the DeCypher (DC), GeneMatcher2 (GM2), 24-processor BlastMachine (BM24) and 12-processor BlastMachibe (BM12). Results are reported for different numbers of input sequences (N) and different alignment types. Missing values indicate that tests were not performed if the general performance trend has already been well established, in order to save many hours of accelerator work.

N	BLAST				HMM			Smith-Waterman		
	DC	BM24	BM12	Grid	DC	GM2	Grid	DC	GM2	Grid
100	74	100	140	30	23	378	28	1 443	160	171
300	96	310	337	68	32	371	39	4 149	366	378
1 000	284	985	1 139	219	92	385	103	14 431	1 032	1 050
3 000	734	2 642	3 356	522	268	400	280	43 250	2 861	2 890
10 000	2 357	8 859	11 122	1 835	861	387	402	n/a	9 442	9 470
30 000	7 057	26 570	33 336	4 670	2 535	398	416	n/a	27 511	27 533
100 000	n/a	n/a	n/a	n/a	9 795	551	570	n/a	n/a	n/a

The improvements provided by the grid load balancing policies are apparent from Table 1. The grid policy for BLAST searches substantially outperformed each of the three available servers for all input sizes, with an approximately 30% speedup over the fastest individual server. The grid policy for HMM was far superior to either of the two available accelerators, approximately matching the performance of the optimal accelerator for each N, while far outperforming the other, suboptimal system. The trivial grid policy for the Smith-Waterman searches, although slightly inferior to a direct job submission to the GeneMatcher2, had an overhead of no more than 30 seconds, which is negligible for jobs that may last several hours. This policy also had the ability to switch to a reserve accelerator automatically if necessary.

These results demonstrate an important advantage of the bioinformatics grid: it can accommodate performance patterns of individual bioinformatics resources that are typically unknown to outside users. The grid also frees the user from the need to resubmit the job to another web service if the previous submission failed. Other advantages of grid systems have been well documented in the computing literature [24, 25]. More sophisticated policies may be designed by expanding the types of information collected by the SGE and accounting for more elaborate scenarios.

We evaluated the Globus® Toolkit interfaces and protocols [26] as an alternative to the commercial Sun N1 Grid Engine. Even though they are considered a *de facto* standard for grid resource access and used in many existing grid systems [27, 28], we discarded the idea. The Globus Toolkit lacks adequate support for semantic job allocation policies, and would have been rather complex to install and configure. Other communication schemas are also possible, such as asynchronous message passing used in [29]. Alternatives to BioMOBY communication mechanism include ^{my}Grid [10], although the two projects have made efforts towards interoperability. A comparison between ^{my}Grid and BioMOBY approaches is presented in [30]. Other systems for dynamic bioinformatics service discovery include [31, 32].

Public access to the Web services involves potential dangers of denial-of-service attacks. None of these problems have thus far been reported for any BioMOBY services to date. Nevertheless, as the BioMOBY usage grows, a BioMOBY-enabled grid system should eventually include the establishment of priorities among clients and detection of patterns of unusual job submissions. Future work should aim at customizing the search with additional parameters once the BioMOBY clients begin supporting this feature, and perfecting the distributed job allocation policy. Propagation of possible failure signals between the grid and BioMOBY systems is another area of interest and a subject of a current Request for Comments #1863: "Exception Reporting in MOBY-S" (available at http://biomoby.open-bio.org).

4 Conclusion

This project has provided unified and transparent access to several high-throughput accelerators for sequence similarity searches. We suggest several load balancing scenarios, which attempt to optimize the deployment of the available computational resources for each type of searches. The details of load balancing between individual accelerators are managed by a grid system. Furthermore, job submission process is standardized via BioMOBY Web services, making the grid completely transparent to users. From the end-user perspective, the project adds several powerful resources that are available for inclusion into BioMOBY data analysis pipelines. Unlike in traditional grid systems, no user accounts are required to utilize our system. We also demonstrate a viable alternative to using the Globus Toolkit. For bioinformatics laboratory administrators, the technical details presented in this report may serve as a practical example of building a wide area bioinformatics grid with minimal installation and user authentication requirements.

Availability of software. Perl scripts and batch-job submission scripts that form the internal data flow mechanism, as well as sample data and other technical instructions, are freely available on the project website http://moby.ucalgary.ca.

Acknowledgments. The authors wish to thank Yingqi Chen and Kathy Laszlo who performed software installation and configuration at the Calgary site. This project is supported by Genome Alberta, in part through Genome Canada, a not-for-profit corporation which is leading a national strategy on genomics with $375 million in funding from the Government of Canada. The infrastructure used in this study is supported by The Alberta Science and Research Authority, Western Economic Diversification, The Alberta Network for Proteomics Innovation, the Canada Foundation for Innovation, Sun Microsystems and iCORE.

References

1. Smith, T.F., Waterman, M.S.: Identification of Common Molecular Subsequences. J Mol Biol. **147** (1981) 195-197
2. Altschul, S.F., Gish, W., Miller, W., Myers, E.W., Lipman, D.J.: Basic Local Alignment Search Tool. J Mol Biol. **215** (1990) 403-410
3. Rabiner, L.R.: A Tutorial on Hidden Markov Models and Selected Applications in Speech Recognition. Proc. of IEEE, **77** (1989) 257-286

4. Gaasterland, T., Sensen, C.W.: Fully Automated Genome Analysis That Reflects User Needs and Preferences: A Detailed Introduction to the MAGPIE System Architecture. Biochimie **78** (1996) 302-310
5. Stein, L.: Creating a Bioinformatics Nation. Nature **417** (2002) 119-120
6. Chicurel, M.: Bioinformatics: Bringing It All Together. Nature **419** (2002) 751, 753, 755
7. Foster, I., Kesselman, C.: The Grid: Blueprint for a New Computing Infrastructure, Morgan Kaufmann: San Francisco, CA (1999)
8. Czajkowski, K., Fitzgerald, S., Foster, I., Kesselman, C.: Grid Information Services for Distributed Resource Sharing. 10th IEEE International Symposium on High Performance Distributed Computing, IEEE Press (2001) 181-184
9. Curbera, F., Duftler, M., Khalaf, R., Mukhi, N., Nagy, W., Weerawarana, S.: Unraveling the Web Services Web - An Introduction to SOAP, WSDL, and UDDI. IEEE Internet Computing 6 (2002) 86-93
10. Stevens, R.D., Robinson, A.J., Goble, C.A.: myGrid: Personalised Bioinformatics on the Information Grid. Bioinformatics. Suppl 1 (2003) i302-i304
11. Goble, C., Stevens, R., Ng, G., Bechhofer, S., Paton, N., Baker, P., Peim M., Brass, A.: Transparent Access to Multiple Bioinformatics Information Sources. IBM Systems Journal 40 (2001) 532–551
12. Hass L., Schwarz, P. M., Kodali, P., Kotlar, E., Rice J.E., Swope W.C.: DiscoveryLink: A System for Integrated Access to Life Sciences Data Sources. IBM Systems Journal 40 (2001) 489–511
13. Foster, I., Kesselman, C., Nick, J., Tuecke, S.: The Physiology of the Grid: an Open Grid Services Architecture for Distributed Systems Integration. Technical report, Global Grid Forum (2002)
14. Wilkinson, M.D., Links, M.: BioMOBY: an Open-source Biological Web Services Proposal. Briefings In Bioinformatics 3 (2002) 331-341
15. Sun N1 Grid Engine 6, http://www.sun.com/software/gridware
16. MOBY Tools, http://mobycentral.icapture.ubc.ca/applets
17. The Common Gateway Interface, http://hoohoo.ncsa.uiuc.edu/cgi
18. GNU Wget, http://www.gnu.org/software/wget
19. BioMOBY in Java, http://biomoby.open-bio.org/CVS_CONTENT/moby-live/Java/docs
20. Oinn, T., Addis, M., Ferris, J., Marvin, D., Senger, M., Greenwood, M., Carver, T., Glover, K., Pocock, M.R., Wipat, A., Li, P.: Taverna: a Tool for the Composition and Enactment of Bioinformatics Workflows. Bioinformatics 20 (2004) 3045-3054
21. Turinsky, A.L., Ah-Seng, A.C., Gordon, P.M.K., Stromer, J.N., Taschuk, M.L., Xu, E.W., Sensen, C.W.: Bioinformatics Visualization and Integration with Open Standards: The Bluejay Genomic Browser. In Silico Biol. 5 (2005) 187-198
22. MOBY Clients, http://biomoby.open-bio.org/index.php/moby-clients
23. National Center for Biotechnology Information, http://www.ncbi.nlm.nih.gov
24. Gentzsch, W.: Grid Computing: A Vendor's Vision. Proc. of CCGrid, (2002) 290-295.
25. Berman, F., Fox, G., Hey, T.: Grid Computing: Making the Global Infrastructure a Reality. Wiley, London (2003)
26. Foster, I., Kesselman, C.: Globus: A Metacomputing Infrastructure Toolkit. Intl J. Supercomputer Applications, 11 (1997) 115-128
27. Frey, J., Tannenbaum, T., Livny, M., Foster, I.T., Tuecke, S.: Condor-G: A Computation Management Agent for Multi-Institutional Grids. Cluster Computing, 5 (2002) 237-246
28. Buyya, R., Abramson, D., Giddy, J.: Nimrod/G: An Architecture for a Resource Management and Scheduling System in a Global Computational Grid. Proc. of HPC ASIA, (2000) 283-289

29. Gannon, D., Bramley, R., Fox, G., Smallen, S., Rossi, Al., Ananthakrishnan, R., Bertrand, F., Chiu, K., Farrellee, M., Govindaraju, M., Krishnan, S., Ramakrishnan, L., Simmhan, Y., Slominski, A., Ma, Y., Olariu, C., Rey-Cenvaz, N.: Programming the Grid: Distributed Software Components, P2P and Grid Web Services for Scientific Applications. J. Cluster Computing. 5 (2002) 325-336

30. Lord, P., Bechhofer, S., Wilkinson, M.D., Schiltz, G., Gessler, D., Hull, D., Goble, C., Stein, L.: Applying Semantic Web Services to Bioinformatics: Experiences Gained, Lessons Learnt. Proc. of 3rd Int. Semantic Web Conference (2004) 350-364

31. Rocco, D., Critchlow, T.: Automatic Discovery and Classification of Bioinformatics Web Sources. Bioinformatics, 19 (2003) 1927–1933

32. Kelly, N., Jithesh, P.V., Simpson, D.R., Donachy, P., Harmer, T.J., Perrott, R.H., Johnston, J., Kerr, P., McCurley, M., McKee, S.: Bioinformatics Data and the Grid: The GeneGrid Data Manager. Proc. of UK e-Science All Hands Meeting (2004) 571-578

Using Public Resource Computing and Systematic Pre-calculation for Large Scale Sequence Analysis

Thomas Rattei[1], Mathias Walter[1], Roland Arnold[2], David P. Anderson[3], and Werner Mewes[1,2]

[1] Department of Genome Oriented Bioinformatics, Technische Universität München, Wissenschaftszentrum Weihenstephan, 85350 Freising, Germany
t.rattei@wzw.tum.de
[2] Institute for Bioinformatics, GSF-National Research Center for Environment and Health, Ingolstädter Landstr. 1, 85764 Neuherberg, Germany
[3] Space Sciences Laboratory, University of California, 7 Gauss Way, Berkeley, CA 94720-7450

Abstract. High volumes of serial computational tasks in bioinformatics, such as homology searches or profile matching, are often executed in distributed environments using classical batching systems like LSF or Sun Grid-Engine. Despite their simple usability they are limited to organizationally owned resources. In contrast to proprietary projects that implement large-scale grids we report on a grid-enabled solution for sequence homology and protein domain searches using BOINC, the Berkeley Open Infrastructure for Network Computing. We demonstrate that BOINC provides a powerful and versatile platform for public resource computing in bioinformatics that makes large-scale pre-calculation of sequence analyses feasible. The FASTA-/Smith-Waterman- and HMMer applications for BOINC are freely available from the authors upon request. Data from SIMAP is publicly available through Web-Services at http://mips.gsf.de/simap.

Keywords: sequence analysis, sequence similarity, protein domain, protein homology, grid computing, public resource computing.

1 Introduction

1.1 Public Resource Computing

Large-scale operations on biological data are typical in bioinformatics. They include sequence similarity searches, analysis of protein domains, protein structure predictions and sequence fragment assembly. Most of these tasks may be executed in small, parallel tasks that do not depend on each other. For instance, sequence similarity searches using heuristics like BLAST [1], FASTA [2] or exact solutions [3,4] may be split into several smaller serial tasks that compare one part of the query sequences with one part of the database sequences.

Classical batching systems, grid computing and public resource computing provide solutions for utilizing existing computer resources to run large-scale serial

W. Dubitzky et al. (Eds.): GCCB 2006, LNBI 4360, pp. 11–18, 2007.

computations. While the classical batching systems like LSF (http://www.platform. com) or the Sun Grid-Engine (http://gridengine.sunsource.net) are focused mostly on UNIX operation systems, the other two paradigms may provide solutions that run on different types of computers. However, there are profound differences between grid computing and public resource computing. The grids consist of organizationally-owned resources like supercomputers, clusters and personal computers. These resources are centrally managed by IT professionals and are connected by high-bandwidth network links. Grid MP is a popular example for organization-level grids. It is a commercial distributed computing software package developed by United Devices. Initiatives like EGEE and EGEE II (http://www.eu-egee.org) aim to develop grid infrastructures that provide access to major computing resources, independent of their geographic location.

In contrast, public resource computing involves an asymmetric relationship between projects and participants. The projects are typically undertaken by small academic research groups with limited compute power. Most participants are independent individuals, who own Windows- or Linux-based PCs or Macintosh computers, connected to the internet by telephone or cable modems. The different preconditions result in different requirements on middleware for public resource computing than for grid computing. For example, features of BOINC (http://boinc. berkeley.edu) like redundant computing, cheat-resistant accounting and support for user-configurable application graphics are not necessary for a grid system. Conversely, grid computing has many requirements that public resource computing does not, such as accommodation of existing commercial and research oriented academic systems and general mechanisms for resource discovery and access.

During recent years, the public resource computing approach has been successfully applied to selected problems in computational biology, like protein structure prediction (Folding@Home, Predictor@Home), protein docking (Rosetta@Home) or epidemiology studies (malariacontrol.net).

1.2 Systematic Pre-calculation of Sequence Analyses

Sequence similarity searches mostly performed by BLAST [1] or FASTA [2] are an essential step in the analysis of any protein sequence and by far the most intensively used bioinformatics methods. Sequence conservation as the basic evolutionary principle implies conservation of structure and function. Thus, structural and functional attributes that cannot be predicted from a single sequence can be efficiently transferred from known to uncharacterized proteins.

The result of any sequence similarity search against a database is a list of significant matches ordered by the similarity score of the pair wise alignments. However, this list represents only a one-dimensional view of the n-dimensional relation between a set of similar and probably evolutionarily conserved sequences. The complete similarity matrix (all-against-all) covers the complete "protein similarity space". Therefore, the information content of an exhaustive database of similarity scores increases substantially since it takes all relations of any similarity sub-graph into account. Employing subsequent analyses such as clustering allows for efficient computation of a number of essential genome analysis tasks applicable to the protein space.

Typically, sequence similarity searches of individual sequences or genomes are repeated frequently since the available datasets change over time. In many analyses this re-computation is the most time consuming step and makes the analysis intractable for large or large numbers of datasets. Therefore, a pre-calculated all-against-all matrix becomes desirable, which stores the similarity space in a database and allows rapid access to significant hits of interest. Our "Similarity Matrix of Proteins" SIMAP [5] maintains a pre-calculated all-against-all matrix representing the protein similarity space and provides comprehensive coverage with respect to the known sequence space.

The non-redundant sequence set of SIMAP is supplemented with protein feature information and cross-references to secondary databases of protein domains and families. The database of associated information is updated automatically whenever new sequences are imported into SIMAP. With the exception of InterPro [6] these features are calculated for the complete set of sequences using an in-house grid computing system. Due to the computationally expensive Hidden Markov Model (HMM) searches for InterPro calculations we import the InterPro hits for all UNIPROT sequences which are provided by the EBI. Additionally we have started to pre-calculate InterPro domains for all sequences that are not yet contained in UNIPROT.

Due to the huge number of sequences and the rapid growth of the public protein databases the calculation of all-against-all protein similarities and InterPro domains for the SIMAP project cannot be performed on typical in-house grid computing systems. Here we describe how we have successfully exploited the public resource computing approach for these calculations.

2 Systems and Methods

BOINC (Berkeley Open Infrastructure for Network Computing) is an open-source platform for public resource distributed computing. BOINC allows a research scientist with moderate computer skills to create and operate a large public resource computing project with about a week of initial work and one hour per week of technical maintenance. The server for the BOINC infrastructure can consist of a single machine configured with common open source software (Linux, Apache, PHP, MySQL, Python).

In our SIMAP system, the computationally expensive tasks of sequence similarity calculations and database searches by profile HMMs are distributed using BOINC. As it was evaluated to be the best compromise between computational speed and sensitivity [7] we have chosen FASTA [2] for finding all putative sequence similarities. The FASTA parameter ktup, substitution matrix and gap penalties are configurable. In order to get exact alignment coordinates and scores of the hits found, every FASTA hit is recalculated using the Smith-Waterman algorithm. The widely used program HMMER (http://hmmer.wustl.edu) has been chosen for profile HMM searches.

3 Implementation

3.1 BOINC Server

Our BOINC server was set up using the BOINC software package on a Linux-based PC server. Apart from the Apache HTTP daemon and the relational database system MySQL which are integral parts of most Linux distributions several BOINC daemon processes needed to be configured:

o the transitioner which implements the redundant computing login was used as provided,
o the validater examines sets of results and was used as provided but may also include application specific result-comparison functions,
o the assimilater handles validated canonical results and was adapted to transfer results into a sequence homology database,
o the file deleter which deletes input and output files from data servers when they are not longer needed was used as provided.

Additionally we have implemented the work generator which creates workunits for the SIMAP project.

3.2 BOINC Applications

The BOINC client consists of two parts: the BOINC core client which is part of the BOINC platform, and the BOINC application which contains the application logic and needs to be implemented for each project.

The BOINC core client is being used by the participants as provided by BOINC. It provides an Application Programming Interface (API) which is a set of C++ functions. The application client is linked to this API and interacts with the core client to report CPU usage and fraction of the workunit completed, to handle requests and to provide heartbeat functionality.

Two applications were implemented in C++. The sequence similarity application uses classes from the FASTA sources to perform sequence similarity calculations resulting in local alignments. The application reads a compressed XML input file which contains the sequences to compare and the parameters that should be used for calculations. The main sections of the input files are:

o FASTA and Smith-Waterman parameters,
o query sequences,
o database sequences,
o a set of test sequences for validation purposes.

The query and database sequence sets are usually of similar length to get the maximal ratio of calculation time per workunit byte. It is not intended to transfer the complete database to the clients, but to distribute quadratic parts of the all-against-all matrix within the workunits. The high number of computing clients makes repeated

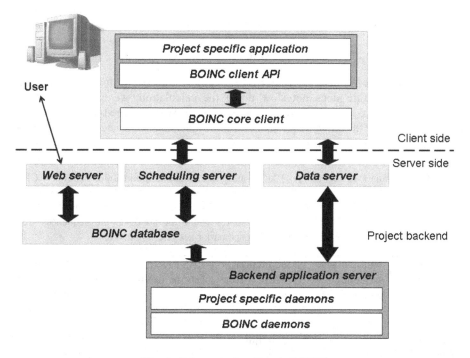

Fig. 1. Schematic data flow in BOINC

transfer of the same database sequences to a certain client very unlikely. Gzip or bzip2 compression ensures minimal size of workunit files.

When the calculation has finished a compressed binary output file is generated which contains all hits found and the all-against-all hits of the test sequence set. Finally the output file is returned by the BOINC core client to our BOINC server. The schematic data flow is shown in Fig. 1.

The application that performs searches for profile HMM matches has been implemented based on the HMMER sources. It gets the sequences and models to compare from a compressed XML input file and works in a similar fashion as the previously described sequence similarity application.

Both applications can also run without BOINC as standalone programs. This feature allows us to use our in-house grid-systems for urgent calculations and to integrate the results into the same workflow as the results from BOINC.

One of the major challenges for the calculation of sequence analyses in a public resource computing system was to derive a ratio between data transfer traffic and calculation time that does not overload the internet connections of the BOINC users. By the combination of compact binary output formats and additional effective input and output data compression we could reduce this ratio to 1.5 Mb download and 500kb upload per hour calculation time on a typical PC.

4 Results

4.1 Performance of BOINC-Applications

To evaluate the performance of our BOINC applications and to make sure that the BOINC infrastructure does not slow down the calculation, we deployed BOINC clients to 10 Linux- and Windows-based PC nodes. We generated 1000 workunits, each containing 2500 query sequences and 2500 database sequences taken from the UNIPROT TrEMBL database.

Table 1. Benchmark results of FASTA[2] binary in standalone mode (Intel Pentium 4 CPU, 2.4 GHz, running Linux), FASTA distributed by GridEngine and the FASTA-based application in BOINC (10 nodes each, Intel Pentium 4 CPU, 2.4 GHz, running Windows or Linux), searching all hits having Smith-Waterman scores >=80 for 5000 sample sequences from UNIPROT-TrEMBL. Average values of 10 benchmark run cycles are given.

Number of workunits	Number of sequence comparisons	Standalone FASTA run time	FASTA in Grid-Engine run time	FASTA application in BOINC run time
1	6.2 million	42min	44min	47min
10	62 million	7h02min	45min	46min
100	620 million	2d20h24min	6h56min	6h54min
1000	6.2 billion	-	2d23h06min	2d22h34min

Comparing the speed of the FASTA binary on a single PC, FASTA distributed using SUN GridEngine and our FASTA based BOINC application we found GridEngine and BOINC to provide similar performance, despite a small overhead in both distributed systems for generating workunits and assimilating results. The run times are given in Table 1. Thus, the BOINC-application performance is equal to that of the program in a classical grid.

4.2 The BOINCSIMAP Project

After extensive in-house tests in order to ensure functionality and stability of the BOINC server and applications, in late 2005 a public BOINC project "BOINCSIMAP" was created that aims to calculate the sequences similarities and InterPro domains for the SIMAP database (http://boinc.bio.wzw.tum.de/boincsimap). In July 2006, the 10.000 users and 20.000 hosts participating in this BOINC project provided an effective computation power of 3.5 TFLOPs to the SIMAP project and dramatically accelerated the calculation of sequence similarities and InterPro domains for new sequences.

During the first months of BOINCSIMAP, several important factors turned out to be crucial for the success of the project. First of all, it was necessary to establish platforms for communication with the participants. These include message boards, announcements and statistics pages, integration of BOINC communities and collaboration with activities propagating BOINC and the BOINC developers. The moderation of the technical and science-related discussions in the message boards is especially time consuming and sometimes difficult with respect to social and cultural differences between the users. A second important success factor is the support for many different hardware and software platforms. Providing application builds for Linux, MacOS or UNIX operating systems, we could integrate users that cannot participate in other BOINC projects that only provide an application for Microsoft Windows. Thus our BOINC team spends much more time on public relations and application development than on the maintenance of the BOINC servers and backend software.

5 Conclusion

We implemented a public resource computing system for sequence similarity searches and profile HMM matching based on BOINC. Most of the software was provided by the open source BOINC project. Only the specific BOINC applications and a number of backend components had to be implemented ourselves. Thus BOINC reduced the amount of time and work necessary for creating and maintaining the large-scale sequence analysis project SIMAP dramatically. The performance per CPU of the BOINC based system is similar to classical batching systems running the corresponding FASTA executable.

The huge numbers of volunteer computers that participate in our BOINC project demonstrate how public resource computing makes large-scale pre-calculation of sequence analyses feasible. Based on this experience we propose the combination of large-scale pre-calculation and public resource computing for related problems in bioinformatics.

6 Availability

The FASTA-/Smith-Waterman- and HMMer clients for BOINC are freely available from the authors upon request. They can be used by other researchers to establish similar BOINC projects for biological sequence analysis.

Sequence similarity and sequence feature data from SIMAP is publicly available through Web-Services at http://mips.gsf.de/simap.

References

1. Altschul, S. F., Gish, W., Miller, W., Myers, G. and Lipman, D. J.: A basic local alignment search tool. J. Mol. Biol., 215 (1990) 403–410.
2. Pearson, W. R.: Flexible sequence similarity searching with the FASTA3 program package. Methods Mol. Biol. 132 (2000) 185-219.
3. Smith, T. and Waterman, M.: Identification of common molecular subsequences. J. Mol. Biol. 147 (1981) 195-197.
4. Needleman, S. and Wunsch, C.: A general method applicable to the search for similarities in the amino acid sequence of two proteins. J. Mol. Biol. 48 (1970) 443- 453.

5. Arnold R, Rattei T, Tischler P, Truong MD, Stumpflen V and Mewes W.: SIMAP--The similarity matrix of proteins. Bioinformatics, 21 Suppl 2 (2005) ii42-ii46.
6. Mulder, N. J., Apweiler, R., Attwood, T.K., Bairoch, A., Bateman, A., Binns, D., Bradley, P., Bork, P., Bucher, P., Cerutti, L., et. al.: InterPro, progress and status in 2005. Nucleic Acids Res. 33 (2005) D201-5.
7. Pearson, W.R.: Searching protein sequence libraries: comparison of the sensitivity and selectivity of the Smith-Waterman and FASTA algorithms. Genomics 11 (1991) 635-650.

Accelerated microRNA-Precursor Detection Using the Smith-Waterman Algorithm on FPGAs

Patrick May[1,2], Gunnar W. Klau[3,4], Markus Bauer[1,5], and Thomas Steinke[2]

[1] Algorithmic Bioinformatics, Institute of Computer Science,
Free University Berlin, Germany
[2] Computer Science Research, Zuse Institute Berlin, Germany
[3] Mathematics in Life Sciences, Institute of Mathematics,
Free University Berlin, Germany
[4] DFG Research Center MATHEON "Mathematics for key technologies",
Berlin, Germany
[5] International Max Planck Research School for Computational Biology
and Scientific Computing, Berlin, Germany
{patrick.may,steinke}@zib.de, {gunnar,mbauer}@mi.fu-berlin.de

Abstract. During the last few years more and more functionalities of RNA have been discovered that were previously thought of being carried out by proteins alone. One of the most striking discoveries was the detection of microRNAs, a class of noncoding RNAs that play an important role in post-transcriptional gene regulation. Large-scale analyses are needed for the still increasingly growing amount of sequence data derived from new experimental technologies. In this paper we present a framework for the detection of the distinctive precursor structure of microRNAS that is based on the well-known Smith-Waterman algorithm. By conducting the computation of the local alignment on a FPGA, we are able to gain a substantial speedup compared to a pure software implementation bringing together supercomputer performance and bioinformatics research. We conducted experiments on real genomic data and we found several new putative hits for microRNA precursor structures.

1 Introduction

One of the most exciting recent discoveries in molecular genetics is the important function of noncoding RNAs (ncRNA) in various catalytic and regulatory processes (see *e.g.* [1,2]). Many experimental studies have shown that large fractions of the transcriptome are consisting on ncRNAs [3,4,5,6]. Additionally, there are several bioinformatics surveys providing evidence for a large amount of ncRNAs in various species [7,8,9]. Due to the ongoing sequencing projects and the large amount of data coming from molecular biology techniques like tiling arrays and cDNA sequencing one of the major challenges in bioinformatics and computational biology will be large-scale analyses and predictions of ncRNAs.

W. Dubitzky et al. (Eds.): GCCB 2006, LNBI 4360, pp. 19–32, 2007.

Such high-throughput, genome-wide searches require fast and efficient detection algorithms as well as high-performance technologies.

MicroRNAs form a conserved, endogenous 21-23 nucleotide class of ncRNAs that is regulating protein-coding gene expression in plants and animals via the RNA silencing machinery (reviewed in [10,11]). Recently, microRNAs have been also discovered in viral genomes indicating that viruses have evolved to regulate both host and viral genes [12]. MicroRNAs are derived from larger precursors that form imperfect ≈ 70 nucleotides long stem-loop structures. In the cytoplasm the precursor is cleaved by the enzyme Dicer to excise the mature microRNA, which is assembled into the RNA silencing machinery. By binding to a complementary target in the mRNA, the mature microRNA inhibits translation or facilitates cleavage of the mRNA [13].

1.1 Previous Work

Computational methods that search for ncRNAs can be divided in two classes: approaches that try to detect ncRNAs *de novo* (like RNAz [14] or QRNA [15]), and programs that search for homologous structures in genomic sequences, given the structure of a known ncRNA family. Programs for the latter task include, *e.g.*, FastR [16] or RSEARCH [17]. FastR takes as its input a so called *filter mask* that basically specifies distance constraints between the various stems. It then searches for exact matching stems of a certain length (the default value is 3) that satisfy these constraints: if all stems are found, the specific region is evaluated again by a more costly alignment algorithm. Since only exact-matching stems are searched, this task can efficiently be done using hash tables, resulting in a very fast filter stage.

RSEARCH on the other hand works with *covariance models*: Covariance models are stochastic context-free grammars that capture both the sequence and the structure of a given training set. Once the parameters of the models are evaluated, it can be used to align the query with a target sequence. RSEARCH provides high specificity and sensitivity at the expense of high computational demands.

Whereas the approaches mentioned above deal with general RNA structures, there are several other approaches that especially focus on the detection of putative microRNA precursor structures. Each one of these, however, relies on additional information in order to increase specificity and sensitivity: RNAmicro [18] depends on a given multiple alignment—which is not available in case of viruses—to detect conserved microRNA precursor structures using essentially the same algorithm as described in [14]. harvester [19] and miralign [20] on the other hand are searching for homologous structures given a set of known microRNA precursors. They use sequence and structure features to classify hairpin structures. miR-abela [21] computes statistical features of hairpin loops and passes these values to a *support vector machine* as its input. Hence, miR-abela relies on known microRNAs to find homologous novel candidates. Furthermore, in order to limit the number of possible hits, the search is restricted to the neighborhood of confirmed hits of microRNAs (the authors reason that microRNAs tend to appear in clusters and the search for novel candidates should therefore be conducted in vicinity to known microRNAs; this assumption is backed up by the computational results conducted in [18]).

1.2 FPGA in Bioinformatics

Field Programmable Gate Arrays (FPGA) are re-configurable data processing devices on which an algorithm is directly mapped to basic processing logic elements, e.g. NAND gates. The clock rate of today's FPGAs is within the range of 100 – 300 MHz and therefore an order of magnitude smaller than those of standard CPUs. To take advantage of using a FPGA one has to implement massively-parallel or data stream oriented algorithms on this re-configurable device. FPGAs are very well suited for the processing of logical and integer data objects represented by flexible— preferably shorter— bit lengths. In this manner, a better silicon utilization compared to standard CPUs with its fixed data representation formats is finally achieved. As a result, FPGA are well suited for certain classes of bioinformatics algorithms. Hence, it is therefore not surprising that FPGA gained a strong attention in the beginning of the genomics era in the 90s. Driven by the exponential growth of sequence data there was and still is an urgent need for faster processing and analysis of genomic data. For some of the standard algorithms in bioinformatics like Smith-Waterman [22] or BLAST [23] commercial black-box solutions are available on proprietary platforms (e.g., from TimeLogic [24]).

Traditionally, the implementation of an algorithm on FPGA, i.e., making a *hardware design*, is accomplished by hardware engineers using hardware description languages like VHDL or Verilog. This approach is known as the "hardware cycle". With the evolution from simple FPGA chips to larger FPGA platforms, FPGA become more and more attractive for application scientists outside the traditional market of embedded systems. Consequently, for this community having no background in hardware design, software tools are available which provide a high-level-language approach *to program* FPGAs (for example, see the C-to-VHDL programming platforms Handel-C, Mitrion-C, or Impulse-C[1]). This high-level programming approach is currently used for our on-going work in implementing bioinformatics application kernels on FPGA.

2   Detecting microRNA Precursor Sequences with Smith-Waterman

This section gives an overview of our computational pipeline called miRo to detect microRNA precursor sequences in genomic sequences. The pipeline consists of several filtering steps, the most important of which employ the Smith-Waterman dynamic programming algorithm for local sequence alignment. Section 2.1 describes how our method works on a single genomic sequence. In

---

[1] Celoxia (www.celoxica.com), Mitrionics Inc. (www.mitrion.com), Impulse Accelerated Technologies Inc. (www.impulsec.com). This list does not claim to be exhaustive.

Sect. 2.2 we show how to use the pipeline in a comparative approach that employs the strong evolutionary conservation of microRNA structures but does not depend on multiple alignments. A detailed description of our method will be given elsewhere [25].

We have implemented miRo in C^{++} as part of the open software library LiSA [26] and report on our computational results in the next section.

2.1   Scanning a Single Genomic Sequence

The main idea of our approach is based on several structural observations about microRNA precursor structures. We define a good microRNA precursor candidate as a subsequence s of the given genomic sequence g with the following characteristics, see also Fig. 1:

1. When folded, s forms essentially one stem and a hairpin loop. Since the two half-stems are in most of the cases not fully complementary, we allow for the formation of interior loops and bulges.
2. The length of the shorter half-stem of s is at least 23 nt in order to accommodate a microRNA.
3. The predicted structure of s should be stable in the sense that folding a sequence with the same nucleotide distribution yields, on average, a much lower minimum free energy.

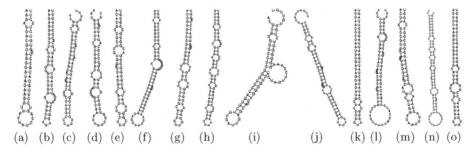

(a)   (b)   (c)   (d)   (e)   (f)   (g)   (h)   (i)   (j)   (k) (l)   (m)   (n) (o)

Fig. 1. Typical microRNA precursor structures detected by our computational pipeline. The picture (generated with RNAfold [27]) shows the first fifteen predicted precursor candidates of the Epstein-Barr virus genome, including known microRNA precursors (a) `ebv-mir-BART17` (miRBase [28] accession id MI0004990), (n) `ebv-mir-BART13` (id MI0003735), and (o) `ebv-mir-BART6` (id MI0003728).

Similar to other approaches, we slide a window of length w along the given genomic sequence g and check whether we detect a good candidate within the window. Folding the subsequence $g[i, \ldots, i + w]$ induced by the window has two major drawbacks:

- Folding algorithms are computationally quite expensive—the most popular algorithms take $O(n^3)$ time, where n is the length of the sequence. Experiments on AMD 2.2 GHz Opteron CPUs showed that processing a sequence of ca. 130.000 nt with a sliding windowsize of 90 nt took about 28 minuteswhen folding the window with RNAfold [27] and only about 1.5 minutes using the Smith-Waterman alignment method described below.
- In general, the *minimum free energy* of a sequence grows with the sequence length. Algorithms based solely on folding must therefore choose the window size very carefully as shorter structures within the window tend to be overlooked.

Since the structure we are looking for is very simple (basically just a stem of a certain minimum length with a hairpin loop), we resort to *local alignment* to approximate the folding process and overcome the two drawbacks mentioned above.

More precisely, consider the window $g[i, \ldots, i + w]$ for some position i on the genome. The main idea is to check whether the subsequence $g[i, \ldots, i + h_{\max}]$ contains a good candidate for a half-stem with a corresponding partner half-stem in $g[i + h_{\min}, \ldots, i + w]$. Here, $h_{\min}$ and $h_{\max}$ are lower and upper bounds on the length of the stems we are looking for. To simulate the folding process we align the reversed sequence $g[i + h, \ldots, i]$ against $g[i + h, \ldots, i + w]$, using a scoring scheme that awards Watson-Crick (A-T, G-C) and wobble base (G-T) pairs and penalizes other pairs and gaps (see Fig. 2 for an illustration of the idea). The scoring matrix used for the SW phases I and II (Fig. 3) of all conducted experiments was: +2 for A-T and G-C, +1 for G-U and -1 for all other nucleotide pairs. The algorithm uses linear gap penalties (-1). We used a window size of 90, a minimal halfstem length of 23 and a maximal halfstem length of 32.

Similar approaches have been used in [29] and [30]. They differ in searching for putative stacks ignoring the connecting regions or searching for microRNAs targets instead of microRNA precursor.

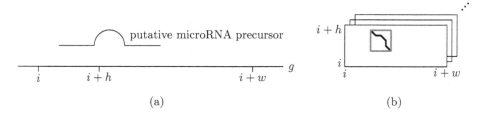

(a)   (b)

Fig. 2. Using local alignment to simulate the folding process (for simplicity, we set $h = h_{\min} = h_{\max}$). (a) Putative microRNA precursor on the genomic sequence g, (b) corresponding alignment.

If the local alignment score is beyond a certain threshold, we include the corresponding subsequence in a list of putative microRNA precursor candidates and apply the following filtering steps, also illustrated in Fig. 3:

1. A window resulting in a good local alignment score is likely to have neighboring windows with similarly good scores. We therefore keep only the best-scoring of overlapping precursor candidates.
2. We have observed that some of the remaining candidates occur more than once on the genomic sequence. We agglomerate these candidates and leave it up to the user whether to keep a representative candidate of the repeated hits or not.
3. We now fold each candidate on the list (currently with RNAfold [27]) and keep those candidates whose predicted structure correspond to a stem-hairpin structure. If this test fails, we inspect the predicted structure and keep a candidate only if the longest stem-hairpin structure satisfies the above criteria. This can be easily done by parsing the bracket notation. Additionally, we discard a candidate if its predicted minimum free energy is above a certain threshold.
   We use an optimized folding parameter set (stack and loop energies) that has been verified experimentally to yield correct microRNA precursor structures.
4. After folding we check whether the longest helical arm is long enough to accommodate a microRNA. If not, we remove the candidate.
5. We compute two z-scores for the remaining candidates. We shuffle the candidate sequence a certain number of times (1000 times for the alignment and 100 times for folding z-score) and compute the alignment score and the minimum free energy of each shuffled sequence. If these z-scores exceed certain thresholds, we discard the corresponding candidate.

2.2 Comparative Application

Applying miRo on a single genomic sequence usually leads to a large list of candidates, which contains still many false positives. This is due to two major reasons:

- It has been verified [31] that large parts of a genomic sequence, *e.g.* there are up to eleven million hairpin structures in the entire human genome, can fold into a microRNA-like structure. Many of these hits are still contained in the final candidate list.
- The list still contains stem-hairpin structures that may be part of a larger non-coding RNA.

We therefore propose a comparative approach (see Fig. 4) to remove most of these false positive hits. The key idea is to apply miRo to k genomes and then compare the candidates in the k output lists against each other. MicroRNA sequences are strongly evolutionary conserved, and we compute a quadratic number of local alignments to detect these similarities. For the SW phase III (Fig. 4

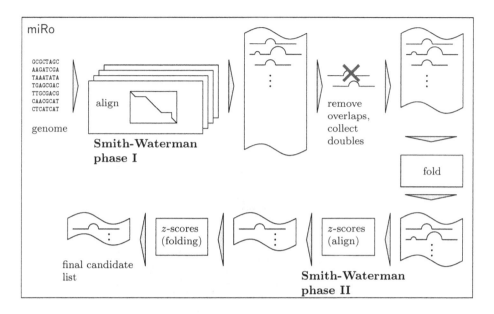

Fig. 3. Overview of the filtering approach on a single genomic sequence

we used an alignment scoring based on matching nucleotides: +1 for identity, -2 for non matching nucleotides, and -2 for gaps.

More formally, we build a *conservation graph* $G_c = (V_c, E_c)$ in the following manner: Each node in V_c corresponds to a microRNA candidate. An edge $(c_1, c_2) \in E_c$ connects two candidates if and only the score of a local alignment between different half-stems is larger than a given threshold. We now enumerate the maximal cliques in G_c and remove candidates from the lists that do not belong to a clique of size m, where m is a user-defined parameter.

We wish to emphasize that our comparative approach, in contrast to other methods, does not rely on good multiple alignments. Often—*e.g.*, in the case of viral genomes—alignments just do not exist, in many other cases, the quality of the multiple alignments is bad due to the difficulty of computing reliable structural alignments.

3   Results

The results presented in this paper show that miRo is able to detect potential microRNA precursors, and, in particular, using an FPGA leads to a considerable computational speedup. A more detailed analysis on specificity and sensitivity will be given elsewhere [25]. In Sect. 3.2 we describe how to build an interface to the FPGA. Our computational results, presented in Sect. 3.3, focus on the acceleration of the method achieved by the use of FPGA technology and support the use of our method for high-throughput genomic experiments.

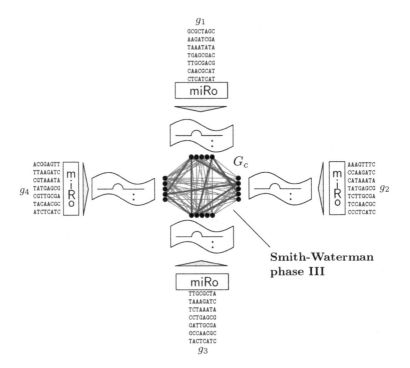

Fig. 4. Overview of the comparative approach for four genomes $g_1, \ldots, g_4$. In the example, the conservation graph G_c contains two maximal cliques of size four and three, respectively.

3.1   Biological Relevance

We tested the miRo algorithm on different problem instances (see Table 1). Epstein, Cytomega, and Human Y represent analyses on single genomes (cf. Fig. 3), whereas Herpes denotes a comparative search on eight different human herpesvirus genomes (cf. Fig. 4).

For the Epstein-Barr virus miRo detected all 23 microRNAs that are currently annotated in the miRbase [28]. Additionally, the tool reported 102 new putative microRNA-precursor structures. After sorting the RNA precursor candidates by their genomic positions we removed those candidates that are overlapping, resulting in 68 candidates. The removed hits possibly point to candidates for other types of ncRNAs, like for instance riboswitches. Since microRNAs tend to build clusters, we removed all candidates that had no neighbor within a 1500kb radius, yielding 52 remaining candidates. The precursors of the 23 known microRNAs can be grouped into four clusters:

1. position $41471 - 43030$ containing 3 microRNAs (*cluster1*),
2. position $139076 - 140107$ containing 8 microRNAs (*cluster2*),
3. position $145932 - 148815$ containing 11 microRNAs (*cluster3*), and
4. position $152745 - 152806$ containing 1 microRNA (*cluster4*).

With the miRo algorithm we can annotate *cluster*1 with one (40974-41042), *cluster*2 with one (140366-140437), *cluster*3 with two (145501-145579, 147160-147230), and *cluster*4 with four (152044-152112, 153755-153819, 154089-154160, 154243-154320) additional microRNA precursor candidates. Moreover, we found eleven new clusters of potential microRNAs precursors: three clusters containing 2 candidates (positions 38335-38477, 72318-73874 and 156845-156944), three clusters containing 3 candidates (positions 12565-13005, 76319-78711 and 90398-90472), three clusters containing 4 candidates (positions 63139-64773, 103513-105494, 150414-150868), one cluster containing 9 candidates (58559-61892), and one cluster containing 10 candidates (47514-55511). A complete list of all clusters and their exact positions will be available for download from the LiSA [26] website.

Within the human cytomegalo virus (*Cytomega*) we found the 10 annotated microRNAs and 200 possible new microRNA precursor candidates, in the sequence of the human chromosome Y (*Human Y*) miRo detected 1294 potential new microRNA precursor structures. Since microRNAs are highly conserved

Table 1. Genomes used in the experiments. For every genome the number of nucleotides, the accession number and a short description is listed.

Problem Case	Size	Genbank Id	Genome Description
Epstein	171823	86261677	Epstein-Barr Herpesvirus 4 wild type
Cytomega	229354	59591	Cytomegalovirus strain AD169
Human Y	6265435	NT_011896.9	Chromosome Y genomic contig
Herpes 1	171657	84519641	Herpesvirus 4 strain GD1
Herpes 2	152261	9629378	Herpesvirus 1
Herpes 3	230287	28373214	Herpesvirus 5 (laboratory strain AD169)
Herpes 4	171823	86261677	Human herpesvirus 4 wild type
Herpes 5	235645	52139181	Human herpesvirus 5 (wild type strain Merlin)
Herpes 6	154746	9629267	Herpesvirus 2
Herpes 7	154746	6572414	Herpes simplex virus type 2 (strain HG52)
Herpes 8	159321	9628290	Herpesvirus 6

across closely related species, we tried to search highly conserved halfstems of microRNA precursor candidates on eight herpesvirus genomes (see Table 1). Therefore, we first searched for potential candidate precursors in all genomes. Totally, we found 127 candidates in *Herpes 1*, 216 candidates in *Herpes 2*, 201 candidates in *Herpes 3*, 131 candidates in *Herpes 4*, 213 candidates in *Herpes 5*, 287 candidates in *Herpes 6*, 292 candidates in *Herpes 7*, and 38 candidates in *Herpes 8*. In the comparative step, in which we perform a search for maximal cliques in the *conservation graph*, we found 4 cliques of size 5 between the genomes *Herpes 1*, *Herpes 2*, *Herpes 4*, *Herpes 6*, and *Herpes 7*. Again, a detailed list of the conserved halfstems can be found on the LiSA [26] homepage.

3.2 Interfacing of the Smith-Waterman Accelerator to LiSA

Software implementation and benchmarks were performed on a Cray XD1 system. Each XD1 node comprises two AMD 2.2 GHz Opteron CPUs with 2 GB host RAM, and an additional FPGA module. This FPGA module consists of a Xilinx Virtex II Pro 50 (speedgrade 7) FPGA attached with four 4 MB QDR SRAM banks. The XD1 system is operated under the Cray release 1.3.

On the Cray XD1 system, we selected the Smith-Waterman Accelerator (SWA) design written in VHDL and kindly provided by Cray Inc. [32,33]. This SWA design implements the basic functionality of the Smith-Waterman dynamic programming algorithm. The scoring function includes the usual terms of a substitution cost table and gap penalties for opening a gap and extending a gap in the sequence substitution.

The Smith-Waterman algorithm runs itself on the FPGA which can be logically viewed as a co-processor. The logic for housekeeping and communication between FPGA and the host node, i.e., the Opteron CPU with its memory must be implemented in a host program. We interfaced the SWA FPGA design to the LiSA [26] library. Based on the core API included in the SWA design [32] a structured API for the LiSA C^{++} implementation was developed. The developed high-level SWA API for LiSA includes the following functionalities:

- *initialization*: allocate the FPGA device and buffer space for the host – FPGA communication, and commit buffer address to the FPGA device.
- *housekeeping of substitution tables*: supports the pre-loading of various 32x32 substitution tables from disk into host memory, scaling of tables entries, and check for potential numerical overflow.
- *preparation*: loads a selected pre-loaded substitution table into the memory window allocated for FPGA – host communication.
- *run*: accepts two sequence strings and the two gap penalties, stores these objects in the memory window for communication, starts the FPGA, and waits for the score result.
- *close*: de-allocates the FPGA device.

The *initialization* and *close* functions has to be called first and last, respectively, if the FPGA is used. A set of functions for *housekeeping of substitution tables* supports the handling of multiple substitution table on the hosts site. With a *preparation* function a selected substitution table is committed to the FPGA device. Any pre-loaded substitution table can be committed before the next call of the *run* function. The *run* function is a high-level abstraction layer to Cray's FPGA library. It includes all the housekeeping functionality for data storage and transfer between host memory and FPGA in context of the Smith-Waterman algorithm.

3.3 Computational Performance

Expected and confirmed by the run-time profile of the prototype version of miRo for the *H*uman Y sequence, see Table 1, two steps dominate the overall computational time: (i) 54% of the time is spent in the calculation of the scores, and (ii) 44% are spent in the folding steps either for folding the candidates or calculating the z-score (see Fig. 3). Therefore, in this paper we focus our work on using FPGA as the accelerator for the calculation of the Smith-Waterman alignment scores.

For three application showcases we report performance numbers. As described above, the calculation of alignment scores is performed with two relatively short sequence strings of length 90. The benchmark measurements were performed on dedicated XD1 nodes allocated by the resource management (batch) system. The timing data are based on *wall-clock times* using the x86 hardware counters[2] Note that measurements based on process timing data (POSIX usr times) make little sense in the context of measuring times where the "I/O" device FPGA is involved.

Table 2 lists the three problem cases and their wall-clock times for the calculation of different scores depending on whether the calculation was performed on the Opteron host CPU ("software") or on the FPGA. The speedup is derived from the ratio of the "software" wall-clock time vs. the time using the FPGA device. The timing for using the FPGA includes overhead times due to the data transfer to and from the FPGA to host memory, respectively. Table 3 summarizes the total runtimes for the three problem cases measured on the Cray XD1 system.

Table 2. Wall-clock times (in seconds) for the Smith-Waterman (SW) miRo score calculations. SW calculations run either on Opteron CPU (marked "software") or VirtexIIPro50 (marked "FPGA") on a Cray XD1 system. The problem cases and tasks are labeled according Table 1 and Fig. 3.

Problem Case	Task	No. of Scores	Software/FPGA	Wall-clock Time	Speedup
Epstein	SW phase I	171733	software	36.29	-
			FPGA	2.62	14.0
	SW phase II	1380000	software	0.23	-
			FPGA	0.03	6.6
Human Y	SW phase I	6265345	software	1337.51	-
			FPGA	94.17	14.20
	SW phase II	1295000	software	0.23	-
			FPGA	0.03	6.6
Herpes	SW phases I	1837144	software	387.86	-
			FPGA	25.86	15.2
	SW phases II	215400000	software	2.38	-
			FPGA	0.30	7.9
	SW phase III	1510570	software	1146.84	-
			FPGA	72.62	15.8

---

[2] The Read Time Stamp Counter (RDTSC) returns the number of clock cycles since the CPU was powered up.

Table 3. Total runtimes (in seconds) for miRo. All calculations have been done either on the Opteron CPU ("Software") or VirtexIIPro50 ("FPGA") on a Cray XD1 system. The problem cases are labeled according Table 1.

Problem Case	Software/FPGA	Wall-clock Time	Speedup
Epstein	software	153.69	-
	FPGA	92.62	1.6
Human Y	software	2458.77	-
	FPGA	955.45	2.6
Herpes	software	3245.68	-
	FPGA	735.36	4.4

The VHDL design of the SWA includes a clock constraint of 100 MHz. Table 2 indicates that even for a rather small strings size of about 100 a respectable speedup of 10 and more can be achieved *including* the data transfer to/from the FPGA. Note, that this translates to a 220+ speedup based on the clock rates of the two compute devices which clearly demonstrates the potential in using re-configurable devices[3].

Due to Amdahl's law, the achieved speedups of 10 and more for the score calculation results finally in total application speedups in the range of 1.7 to 4.4, depending on the problem case. At the first glance these speedup numbers may appear rather small compared to results reported elsewhere but we emphasize that the search for miRNA substructures involves only short sequences and the potential speedup on the FPGA is reduced due to the additional calling and data transfer overhead, e.g., see the speedup of about (only) 8 for the calculation of a large number of z-scores in the *Herpes* problem case.

4   Summary and Conclusion

We have introduced our framework miRo for the detection of microRNA precursor structures based on the Smith-Waterman algorithm. We conducted experiments on real genomic data and we found several new putative hits for microRNA precursors. Furthermore, the miRo implementation interfaced to a Smith-Waterman Accelerator design on a XC2VP50 FPGA at 100 MHz gives *total application speedups* of 4 for the calculation of Smith-Waterman scores for reasonable but still small sized problems. Thus, this implementation is a step forward towards the high-throughput analysis of RNA sequences and also gives an example how recent hardware technologies and bioinformatics research can benefit from each other.

Acknowledgments. We would like to thank Jenny Russ for implementing an early version of the miRo idea, and Christian Köberle for sharing his knowledge about microRNAs with us.

---

[3] Note, that for other problem cases with larger sequence strings the same SWA design demonstrated even larger speedups of 28x. [33].

References

1. J. S. Mattick. Challenging the dogma: the hidden layer of non-protein rnas in complex organisms. *Bioessays*, 25:930–939, 2003.
2. J. S. Mattick. RNA regulation: a new genetics? *Nature Genetics*, 5:316–323, 2004.
3. D. Kampa *et al.* Novel RNAs identified from an in-depth analysis of the transcriptome of human chromosomes 21 and 22. *Genome Res.*, 14(3):331–42, 2004.
4. J. M. Johnson, S. Edwards, D. Shoemaker, and E. E. Schadt. Dark matter in the genome: evidence of widespread transcription detected by microarray tiling experiments. *Trends Genet.*, 21(2):93–102, 2005.
5. T. Imanishi *et al.* Integrative annotation of 21,037 human genes validated by full-length cDNA clones. *PLos Biology*, 2:856–875, 2004.
6. J. M. Cummins *et al.* The colorectal microRNAome. *PNAS*, 103(10):3687–3692, 2006.
7. S. Washietl, I. L. Hofacker, M. Lukasser, A. Huttenhofer, and P. F. Stadler. Mapping of conserved RNA secondary structures predicts thousands of functional noncoding RNAs in the human genome. *Nat Biotechnol.*, 23(11):1383–1390, 2005.
8. K. Missal, D. Rose, and P. F. Stadler. Non-coding RNAs in ciona intestinalis. *Bioinformatics*, 21(S2):i77–i78, 2005.
9. K. Missal *et al.* Prediction of structured non-coding rnas in the genomes of the nematodes caenorhabditis elegans and caenorhabditis briggsae. *J. Exp. Zoolog. B. Mol. Dev. Evol.*, page Epub ahead of print, 2006.
10. V. Ambros. The functions of animal microRNAs. *Nature*, 431:350–355, 2004.
11. D. P. Bartel. MicroRNAs: genomics, biogenesis, mechanism, and function. *Cell*, 116:281–297, 2004.
12. C. S. Sullivan and D. Ganem. MicroRNAs and viral infection. *Cell*, 20:3–7, 2005.
13. L. He and G. Hannon. MicroRNAs: small RNAs with a big role in gene regulation. *Nat. Rev. Genet.*, 5:522–531, 2004.
14. Stefan Washietl, Ivo L. Hofacker, and Peter F. Stadler. Fast and reliable prediction of noncoding RNAs. *PNAS*, 102(7):2454–2459, 2005.
15. E Rivas and S R Eddy. Noncoding RNA gene detection using comparative sequence analysis. *BMC Bioinformatics*, 2, 2001.
16. S. Zhang, B. Haas, E. Eskin, and V. Bafna. Searching genomes for noncoding RNA using FastR. *IEEE/ACM Trans. Comput. Biology Bioinform.*, 2(4):366–379, 2005.
17. R. Klein and S. Eddy. RSEARCH: finding homologs of single structured RNA sequences. *BMC Bioinformatics*, 4, 2003.
18. J. Hertel and P. Stadler. Hairpins in a haystack: Recognizing microRNA precursors in comparative genomics data. In *ISMB'06*, 2006. To appear.
19. T. Dezulian, M. Remmert, J. Palatnik, D. Weigel, and D. Huson. Identification of plant microRNA homologs. *Bioinformatics*, 22(3):359–360, 2006.
20. X. Wang, J. Zhang, F. Li, J. Gu, T. He, X. Zhang, and Y. Li. MicroRNA identification based on sequence and structure alignment. *Bioinformatics*, 21(18):3610–3614, 2005.
21. A. Sewer, N. Paul, P. Landgraf, A. Aravin, S. Pfeffer, M. Brownstein, T. Tuschl, E. van Nimwegen, and M. Zavolan. Identification of clustered microRNAs using an ab initio prediction method. *BMC Bioinformatics*, 6(1):267, 2005.
22. T. F. Smith and M. S. Waterman. Identification of common molecular subsequences. *J. Mol. Biology*, 147:195–197, 1981.
23. S. F. Altschul, W. Gish, W. Miller, E. W. Myers, and D. J. Lipman. Basic local alignment search tool. *J. Mol. Biol.*, 215:403–410, 1990.

24. Active Motif Inc. TimeLogic DeCypher solutions. `http://www.timelogic.com/decypher_algorithms.html`.

25. P. May, M. Bauer, C. Koeberle, and G. W. Klau. A computational approach to microRNA detection. Technical Report, Zuse Institute Berlin, 06-44, 2006.

26. LiSA—Library for Structural Alignment. `http://www.planet-lisa.net`.

27. Ivo L. Hofacker. Vienna RNA secondary structure server. *Nucl. Acids Res.*, 31(13):3429–3431, 2003.

28. S. Griffiths-Jones, R. J. Grocock, S. van Dongen, A. Bateman, and A. J. Enright. miRBase: microRNA sequences, targets and gene nomenclature. *Nucleic Acids Research*, 34(Database Issue):D140–D144, 2006.

29. A. J. Enright, B. John, U. Gaul, T. Tuschl, C. Sander, and D. S. Marks. MicroRNA targets in *Drosophila*. *Genome Biology*, 5(1):R1.1–R1.14, 2003.

30. V. Bafna, H. Tang, and S. Zhang. Consensus folding of unaligned RNA sequences revisited. *J. Comput. Biol.*, 13(2):283–295, 2006.

31. I. Bentwich, A. Avniel, Y. Karov, R. Aharonov, S. Gilad, O. Barad, A. Barzilai, P. Einat, U. Einav, E. Meiri, E. Sharon, Y. Spector, and Z. Bentwich. Identification of hundreds of conserved and nonconserved human microRNAs. *Nature Genetics*, 37(7):766–770, 2005.

32. S. Margerm. Cray XD1 Smith Waterman Accelerator (SWA) FPGA Design. PNR-DD-0025 Issue 0.7, Cray inc., December 2005.

33. S. Margerm. Reconfigurable computing in real-world applications. *FPGA and Structured ASIC Journal*, February 2006. `http://www.fpgajournal.com/articles_2006/pdf/20060207_cray.pdf`.

Implementation of a Distributed Architecture for Managing Collection and Dissemination of Data for Fetal Alcohol Spectrum Disorders Research

Andrew Arenson[1], Ludmila Bakhireva[2], Tina Chambers[2], Christina Deximo[1],
Tatiana Foroud[3], Joseph Jacobson[4], Sandra Jacobson[4], Kenneth Lyons Jones[2],
Sarah Mattson[5], Philip May[6], Elizabeth Moore[7], Kimberly Ogle[5], Edward Riley[5],
Luther Robinson[8], Jeffrey Rogers[1], Ann Streissguth[9], Michel Tavares[1],
Joseph Urbanski[3], Helen Yezerets[1], and Craig A. Stewart[10]

[1] Indiana University, University Information Technology Services, Indianapolis,
IN 46202, USA
{aarenson, cdeximo, jlrogers, mtavares, yyezert}@indiana.edu
[2] University of California, San Diego, Department of Pediatrics, La Jolla, CA 92093, USA
{lbakhireva, klyons, chchambers}@ucsd.edu
[3] Indiana University School of Medicine, Department of Medical and Molecular Genetics,
Indianapolis, IN 46202, USA
{tforoud, joaurban}@iupui.edu
[4] Wayne State University, Department of Psychiatry and Behavioral Neurosciences, Detroit,
Michigan, USA
{joseph.jacobson, sandra.jacobson}@wayne.edu
[5] San Diego State University, Center for Behavioral Teratology, San Diego, CA 92120, USA
{smattson, kowens, eriley}@sdsu.edu
[6] University of New Mexico, Center on Alcoholism, Substance Abuse & Addictions,
Albuquerque, NM 87106, USA
pmay@unm.edu
[7] St. Vincent's Hospital, Indianapolis, IN 46032, USA
ESMoore@stvincent.org
[8] State University of New York, Buffalo, New York 14260, USA
lutherkr@buffalo.edu
[9] University of Washington Medical School, Department of Psychiatry and Behavioral
Sciences, Fetal Alcohol and Drug Unit, Seattle, Washington 98195, USA
astreiss@u.washington.edu
[10] Indiana University, Office of the Vice President for Information Technology,
Bloomington, IN 47405, USA
stewart@iu.edu

Abstract. We implemented a distributed system for management of data for an international collaboration studying Fetal Alcohol Spectrum Disorders (FASD). Subject privacy was protected, researchers without dependable Internet access were accommodated, and researchers' data were shared globally. Data dictionaries codified the nature of the data being integrated, data compliance was assured through multiple consistency checks, and recovery systems provided a secure, robust, persistent repository. The system enabled new types of science to be done, using distributed technologies that are expedient for current needs while taking useful steps towards integrating the system in a future grid-based cyberinfrastructure. The distributed architecture, verification steps, and data dictionaries suggest general strategies for researchers involved

W. Dubitzky et al. (Eds.): GCCB 2006, LNBI 4360, pp. 33–44, 2007.
© Springer-Verlag Berlin Heidelberg 2007

in collaborative studies, particularly where data must be de-identified before being shared. The system met both the collaboration's needs and the NIH Roadmap's goal of wide access to databases that are robust and adaptable to researchers' needs.

Keywords: Distributed Computing, Repository, Data Dictionary, Fetal Alcohol Spectrum Disorders.

1 Introduction

This paper describes the information technology infrastructure required to support distributed research on uncommon diseases such as those represented by FASD – an infrastructure which allowed researchers to collaboratively create and use a common, pooled data resource to enable discoveries and insights that would otherwise not be possible. FASD refers to a range of debilitating effects on the central nervous system in children who were exposed to alcohol as a fetus as a result of maternal alcohol consumption [1]. A variety of FASD diagnoses, such as Fetal Alcohol Syndrome (FAS), Fetal Alcohol Effects (FAE), and Alcohol-Related Neurodevelopmental Disorder (ARND) are differentiated primarily by the presence or absence of facial features, but all share developmental abnormalities in adaptive functioning, attention and memory problems, distractability, learning problems, poor judgment, and fine and gross motor difficulties [2]. Fetal Alcohol Spectrum Disorders are entirely preventable through maternal abstinence; yet an estimated 1-2 out of 1000 children per year in the United States are born with FAS, the most severe form of FASD [3]. One of the roadblocks to research is obtaining significantly-sized populations to study. The research challenge presented by the relatively low incidence of FASD is magnified by difficulties in physical diagnosis as distinguishing features may depend upon race and age. Only a small number of experts are currently qualified to make a definitive diagnosis of FASD. The National Institute on Alcohol Abuse and Alcoholism (NIAAA, one of the National Institutes of Health) has funded an international research consortium called the Collaborative Initiative on Fetal Alcohol Spectrum Disorders (CIFASD) to accelerate research in FASD and in particular to create new ways to make authoritative, differential diagnoses of FASD. CIFASD has as one of its fundamental organizing principles that more rapid progress can be made in understanding and developing interventions for FASD by combining the efforts of multiple researchers at multiple sites.

1.1 Distributed Nature of CIFASD

The Collaborative Initiative on Fetal Alcohol Spectrum Disorders consists of researchers and study populations from six countries. Some of the participating sites are in relatively remote areas with little access to the Internet. The privacy of the populations must also be protected, including careful management of subject identifying information and data which are inherently identifiable (e.g. facial images). The combination of worldwide collection of data and the need to protect subject privacy led to a data management strategy based on a distributed architecture that enabled local entry and management of data along with subsequent submission and sharing of those data via a central repository.

1.2 Commitment to Data Integration

Because CIFASD includes research groups that had previously been operating independently, the consortium made a joint commitment to use common data definitions and data dictionaries in order to make it possible to integrate data from different sites. To enable secure and reliable storage of data and to make sharing and analysis of data across multiple sites and multiple research groups possible, CIFASD also agreed that shared consortium data would be stored in a central repository. A fundamental design feature agreed to by CIFASD was that data would be added to the central data repository on a record-by-record basis. A record typically consists of data from a single modality (e.g. test result, interview response, or image) for an individual child or parent. Records were allowed into the central repository only when fully in compliance with the data dictionary. Thus, another design constraint was building a mechanism that ensured that all records added to the central repository be unique and in compliance with the data dictionary.

1.3 Clinical Needs vs. Computer Science Research

The tools required by the collaboration were built by the Informatics Core of CIFASD in close cooperation with members of projects and other cores from across the collaboration. There is an inherent conflict between meeting the immediate requirements of end users quickly versus allowing computer science collaborators to create innovative tools to meet those demands [4]. We dealt with this conflict by expediting an initial rollout of software using well-understood technology, while laying the groundwork to migrate the software in the future to use advanced cyberinfrastructure techniques.

2 System Design

The practical requirements dictated by the nature of the consortium, distributed data collection, and data analysis objectives called for a two-tier system enabling local data entry (at times no connection to the Internet was possible), a central data repository, and a mechanism for verifying and uploading data to the central repository.

2.1 Data Dictionaries

The creation of data dictionaries used across the consortium was critical to its overall success. Earlier attempts to analyze data across multiple research studies encountered problems due to the difficulties of integrating data from multiple sites. These problems included different definitions of terminology and different experimental designs that led to tantalizingly similar but not quite comparable datasets. The details of the data dictionaries and their content will be described elsewhere. Data dictionaries were created for modalities including dysmorphology, neurobehavior, prenatal alcohol consumption, 3D facial images, and control factors. For these different sources of data the number of variables per modality ranged from 20 to greater than 800 variables. Data types included numeric, category, and text data as well as 2D and 3D images.

One of the most difficult challenges involved in creating data dictionaries for the collaboration involved reaching consensus on which measures to use for prenatal alcohol consumption. Issues included cultural differences between populations and an interest in providing high quality tools for future data capture while still incorporating data from pre-existing studies. Cultural differences include such things as: differences in interpretations as to the quantity of alcohol in a single drink or how much alcohol consumption is required for a 'binge'; differences in the types of alcohol consumed; and differences in the best ways to elicit accurate responses from interviewees. A committee of representatives from the projects in the collaboration were brought together to determine a core set of variables that could be generated from the various instruments used to capture prenatal alcohol consumption, including agreeing on definitions and valid ranges for these variables. The core data dictionary was designed for use by researchers with pre-existing data. At the same time an expanded data dictionary was created to encompass the tenfold larger set of variables that had been used by two of the collaboration's projects to capture the raw data required to calculate the core variables. This expanded data dictionary was designated as the standard for any new projects entering the collaboration. Because of the possibility that new projects might not be able to use the expanded data dictionary due to cultural differences, such projects would fall back to using only the core variables if necessary. In this way the widest range of research was supported, allowing comparisons of the smaller set of core variables across all projects while still allowing analysis of the wider range of raw variables for those projects which used the expanded alcohol and control data dictionary. Thus we avoided the problem noted in other fields such as microarray analysis [5] where competing standards prevent data from different sites from being integrated.

It is unrealistic to expect any requirements gathering process to anticipate all future needs of researchers. It has been shown, for instance, that combining coded fields with free text fields provides the fullest assessment of a clinical database [6]. Free text fields were provided to enhance the effectiveness of the data collected by the collaboration and later proved useful in understanding seeming discrepancies in the data that had been entered.

Similarly, the creation of data dictionaries proved to be an ongoing rather than onetime process. The data dictionaries' primary importance lay in forcing and formalizing agreement on precisely what data were captured so that the data could be integrated. The nature of the data being captured, however, changed over time. For the most part, required changes to the data dictionary involved extensions for new types of data being captured. At times, however, experience of users attempting to capture or analyze data led to uncovering differences in interpretations of the data dictionary that necessitated refinements or corrections. Further it is to be expected, as has been seen in other research [7], that clinical practice will change over time, requiring data dictionaries, software tools, and repositories to change as well.

One issue that arose on a regular basis was the need to clarify the allowable ranges for various measures. When a subject scored a value outside of the allowable range for a particular variable, the designated authorities from within the collaboration were called upon to work with the data manager to determine whether the instrument was administered correctly or the allowable ranges needed to be expanded in the data dictionary.

2.2 Data Entry

Individual projects designated data managers, responsible for working with the CIFASD Informatics Core to assure that data was submitted to the central repository.

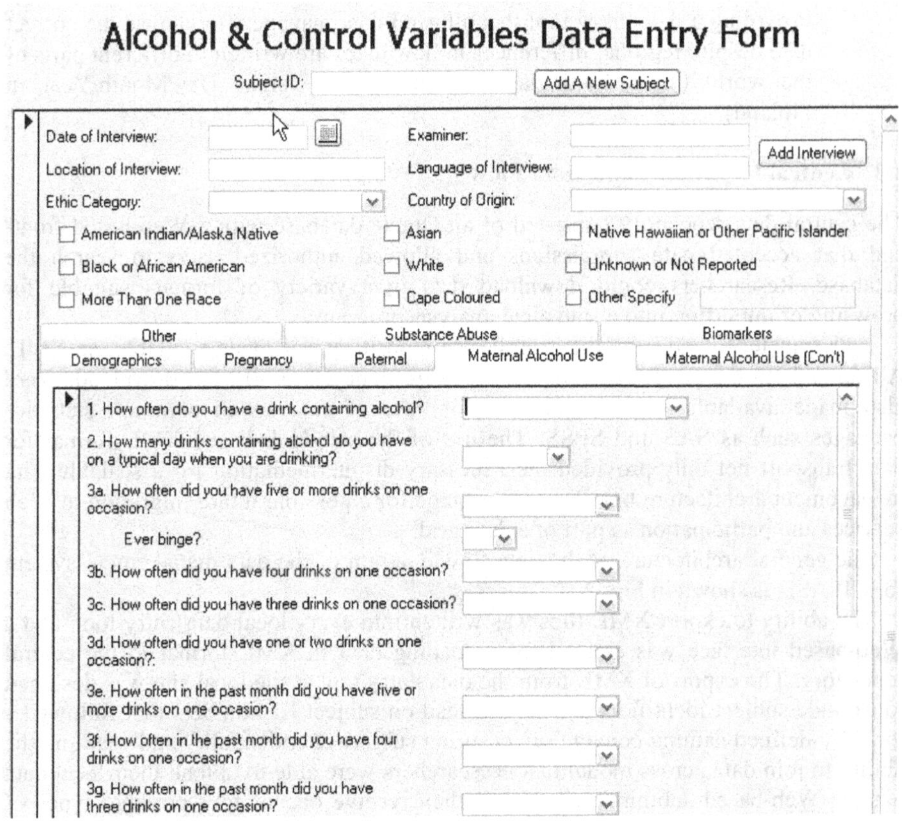

Fig. 1. Screenshot of the beginning of one section of the data entry GUI for Alcohol & Control variables. This GUI entry tool holds over 250 unique variables which can be exported as XML and transferred to the CIFASD central repository. Note the use of tabs (*middle*) to allow data managers to quickly maneuver amongst subsections of the data to be entered.

Data managers were sometimes, but not always, the same people who were responsible for data entry. Data entry for the text and numeric data was accomplished through the use of Microsoft Access® databases, running standalone on PCs. Microsoft Access® was selected because of the combination of the functionality it provides for data entry and its wide availability to researchers around the world, most of whom already had the software installed. Separate graphical user interfaces (GUIs) were created for each modality (Dysmorphology, Neurobehavior, and Alcohol & Control) that required manual data entry. See Fig. 1 for an example. The GUIs proved efficient for handling the common data entry tasks of providing drop-down menus for category choices, checking data types, and checking data ranges. The GUIs also helped solve less common problems as they arose, such as:

- Providing a series of tabs that allowed data managers to navigate quickly amongst different subsections of a dataset to be entered
- Caching metadata to be applied to some but not all of a series of subtests
- Providing a date entry tool that allowed data managers to choose the correct date despite regional differences in how dates are written in different parts of the world (e.g. Month/Day/Year in the US versus Day/Month/Year in Finland)

2.3 Central Repository and Data Flow

The central data repository consisted of an Oracle database with a Web-based front-end that accepted data submissions and allowed authorized users to search the database. Researchers could download data in a variety of formats suitable for browsing or importing into a statistical analysis program.

Data transfer to and from the central data repository was made available via XML. All uploads to the central data repository were done via XML, but downloads were also made available in other formats for ease of analysis in common statistics packages such as SAS and SPSS. The use of the standards-based XML format for data transport not only provided the necessary disintermediation for a scalable data management architecture but also set the stage for a possible future migration to Web services and participation as part of a data grid.

The general architecture of the data flow diagram of the data management system for CIFASD is shown in Fig. 2.

The ability to export XML files was written into every local data entry tool, and a Web-based interface was created for uploading data in XML format to the central repository. The export of XML from the data entry tool at the local site was designed to exclude subject identifiers and rely instead on subject ID numbers that followed a globally-defined naming convention, ensuring subject confidentiality and enabling the ability to join data across modalities. Researchers were able to upload their local data via the Web-based submission tool and then receive one of four possible types of feedback, as follows:

- Data were compliant with the data dictionary, did not match any existing records, and were thus imported into the central data repository.
- Data were in some way incomplete or noncompliant with the data dictionaries and were thus excluded from entering the central repository.
- Data were an exact duplicate of existing data in the central repository, and thus no action needed to be taken, as the central repository was already up to date with respect to those data.
- Data appeared to be an update to an existing record. The data manager was then given an opportunity to review differences between the existing central repository version and the attempted upload and then queried as to whether or not the new record should have been considered an update of the prior record. New data identified as updates by the data manager were then applied to the central repository. Data not identified as a valid update by the data manager were ignored.

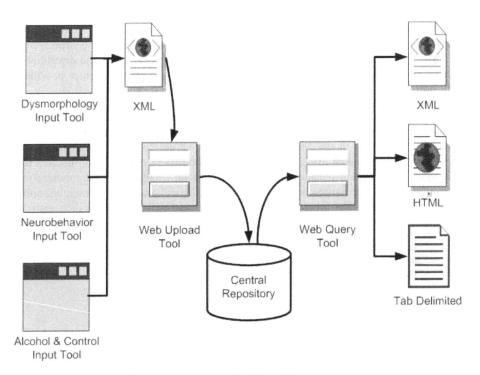

Fig. 2. Diagram of data flow architecture for CIFASD. GUI-based data entry tools used to store data at local sites (*left*) produced XML-formatted, de-identified data, which was uploaded to the central repository and could be retrieved in a variety of formats (*right*).

The distributed nature of the CIFASD architecture also allowed for flexibility in how different types of data were handled or in what tools local sites used for data entry. The 3D Facial Imaging Core, for instance, collected data that were generated from software and did not involve data entry. These data were still transferred to the central repository using a Web-based XML submission. Similarly, one of the local sites collected data using scanning technology and optical character recognition and did not require a separate data entry tool. Their data were accommodated by translating the results of their data collection process to XML before using the Web-based XML submission.

The data collected by the collaboration were complex in terms of representing a broad range of modalities: 2D and 3D facial images, brain images, facial examinations, neurobehavioral instruments, and questionnaires for prenatal alcohol consumption and control factors. In terms of the structure of the XML encoding and the central repository's relational database, much of this complexity was reduced through segregating by modality the variables and images that were to be collected and allowing data across modalities to be joined using the globally unique subject identifiers. For example, the dysmorphology data and the neurobehavior data had different concepts of uniqueness. Dysmorphology defined a unique record as being a unique tuple of subject id, examiner, and examination date while neurobehavior allowed a subject to have any subset of eighteen subexaminations but to not have any subexamination represented more than once. By grouping data first by modality, the peculiar needs of each modality could be handled individually within the structure of the XML file, within the code in the upload tool that parsed the XML and determined whether or not the data were compliant, and within the query tools that provided XML output to researchers.

Example of XML showing how complexity was handled via segregating variables by modality. Similar measures attained from 3D facial images and dysmorphological exams are combined in the same dataset. (Note: These data reflect that the same measurements were taken using different units in the two modalities – millimeters for the computer-calculated measurements from 3D facial imaging and centimeters for the hand-calculated measurements from dysmorphological exams. Measurement units and allowable ranges for all measures were strictly defined in the data dictionaries, but could vary between modalities.)

```
<Global ID="SMS43">
 <FacialImaging>
  <Scan Date="13-SEP-2005">
   <PhiltrumLength>14.77</PhiltrumLength>
   <RightPalpebral>28.02</RightPalpebral>
   <LeftPalpebral>28.06</LeftPalpebral>
   etc.
  </Scan Date>
 </FacialImaging>
 <Dysmorphology>
  <ExaminerLastName>Jones</ExaminerLastName>
  <DateExam>2005-08-16T00:00:00</DateExam>
  <PhiltrumLength>1.4</PhiltrumLength>
  <PFL>2.8</PFL>
  etc.
 </Dysmorphology>
</Global ID>
```

2.4 Assuring Data Quality

The data dictionary's upfront purpose of forcing a shared understanding of the meaning of the data being collected was followed by the data dictionaries being used to ensure that only appropriate data were collected. This was implemented in three layers. The two primary layers were at the point of data entry and the point of data submission. The data entry tools included type and range checking as well as dependency checks between variables. The data submission tools repeated the type, range, and dependency checks both to accommodate the modalities and local sites that did not use a data entry tool, but also to safeguard against instances where the data could have been accidentally changed after data entry. The third layer of protection based on the data dictionary came from a check that can be run from within the data entry tool independent of data entry itself. Although unlikely, it was possible for researchers to bypass the type, range, and consistency checks during data entry, so the independent check allowed data managers to check their data locally to make sure that there were no inconsistencies before performing an export to XML and submission to the central repository.

3 Reliable Infrastructure

The data in the central repository were safeguarded via a variety of mechanisms. Access to the central data repository was controlled by authentication via a local Web-based authentication method – Indiana University's Central Authentication Service (CAS) [8]. This provided a ready way both to provision access to Indiana University Web servers by non-Indiana University personnel via its support for user-creatable Guest Accounts and to integrate with the Web-based applications used for data submission and retrieval. The servers that host the system were maintained by professional staff in modern computing facilities and were subject to regular security scans by a security group that operates independently of the system administration groups. The data resided on disk storage systems with built-in redundancy to protect against the failure of any single disk. The data were backed up to the Indiana University Massive Data Storage System [9], which uses High Performance Storage System (HPSS) software [10] to write data simultaneously to automated tape libraries in two geographically distributed computing facilities (one in Indianapolis, IN; the other in Bloomington, IN). These automated tape libraries – and the Indiana University data centers in Indianapolis and Bloomington – are connected via the University-owned I-Light network [11]. This provided reliable storage of data and resilience against any sort of natural or manmade disaster short of a regional disaster affecting both data centers simultaneously.

4 Results and Conclusions

The data management facilities created for the Collaborative Initiative on Fetal Alcohol Spectrum Disorders proved effective in supporting the needs of the collaboration. The data integrity functionality embedded throughout the data management facilities provided the technical compliance assurance required for effective integration and analysis of the data collected by the members of the CIFASD. The distributed architecture of the system enabled researchers to collect data locally, maintain subject privacy, work in areas without robust Internet access, and still share data globally with the rest of the collaboration. Researchers were able to aggregate results from multiple populations for joint analysis in ways that were not previously possible and with a high degree of assurance in the quality of the data in the CIFASD central repository.

The critical proof of the data management infrastructure was in the quantity of subjects included and in the new types of science that were enabled. The CIFASD central repository accrued records for 224 subjects with FAS and as many as 463 that had at least some symptoms of FAS and might eventually be diagnosed as FASD. In comparison, one of the largest previous studies included only 88 FAS subjects [12]. Researchers were also able to analyze both cross-population datasets [13] and cross-modality datasets [14] on a scale that had never before been achieved. The increase in numbers of affected subjects provided a concomitant increase in the power of statistical analysis to detect differences amongst different subsets of the affected. The aggregation of results from many populations increased the range of possible hypotheses that could be examined, including comparisons amongst ethnicities and comparisons including some demographic and alcohol consumption categories that reached significant numbers only through inclusion of subjects from multiple sites.

The CIFASD data repository proved to be a practical, distributed system that leveraged current technology to provide new capabilities for important medical research. The most difficult operational challenges were in reaching consensus on the data dictionaries. The development of the CIFASD applications required some cleverness to handle the secure management of sensitive data, to manage data uploads and retrieval by a group of people with highly variable access to the Internet, and to resolve some issues that arose because the collaboration was international in scope (e.g. the matter of date formats). The use of XML for data transfer and the reliance on a commonly available commercial database product for data entry were key elements in the rapid implementation of a system that successfully addressed the needs of the CIFASD researchers.

The total amount of data stored in the central data repository was relatively modest - roughly 2 GB. This should grow substantially over time. In the long run, we plan to migrate the CIFASD data management process and central repository to the NSF-funded TeraGrid [15]. One of the goals of the TeraGrid is to improve the general efficiency of scientific research within the US, and Indiana University is deeply involved in the TeraGrid as a Resource Provider. While the data management and analysis needs of the collaboration could easily be handled within Indiana University's facilities, using the infrastructure of the TeraGrid will enable better scalability and robustness of IU's services. Primary benefits will include leveraging the authentication and authorization infrastructure of the TeraGrid and increasing the disaster resilience of the CIFASD data repository. Copies of data are already stored in multiple locations within the Midwest region of the United States, but via the TeraGrid, it will be straightforward to also maintain backup copies of the data repository at locations that would remain safe even in the event of a regional disaster. Most importantly, leveraging the nationally-funded TeraGrid effort allows for economy of scale in the creation and management of cyberinfrastructure rather than crafting a new solution for each collaborative project.

The distributed data management architecture created in support of CIFASD served its purpose of supporting research and discovery related to the study of Fetal Alcohol Spectrum Disorders and is a useful model for any collaboration dealing with the needs of geographically separated sites that need to protect patients' privacy, deal with sometimes uncertain Internet access, and still allow researchers to share data. Further, the consortium created a persistent repository of interest that was and continues to be useful to studies of fetal alcohol spectrum disorders or related health impacts of maternal alcohol consumption. This persistent resource enabled more researchers to get at a wider set of important data and helped the consortium achieve the general NIH goal of "wide access to technologies, databases and other scientific resources that are more sensitive, more robust, and more easily adaptable to researchers' individual needs" [16]. The implementation of the CIFASD data management architecture was an effective use of distributed computing in support of a particular research project, which led to science results that would otherwise have been difficult or impossible to enable.

Acknowledgments. This research was supported in part by the Informatics Core for the Collaborative Initiative on Fetal Alcohol Spectrum Disorders, which received support from the National Institute on Alcohol Abuse and Alcoholism under grant 1U24AA014818. This research was also supported in part by the Indiana Genomics Initiative. The Indiana Genomics Initiative is supported in part by Lilly Endowment, Inc. The facilities that housed the CIFASD data repository were created with the assistance of grants-in-kind of hardware equipment from STK, Inc. (now a subsidiary of Sun Microsystems, Inc.) and Shared University Research grants from IBM, Inc. IU's participation in the TeraGrid is funded by National Science Foundation grant numbers 0338618, 0504075, and 0451237.

References

1. Barr HM, Streissguth AP: Identifying maternal self-reported alcohol use associated with fetal alcohol spectrum disorders. Alcohol Clin Exp Res. 2001;25:283-7.
2. Streissguth AP, O'Malley K. Neuropsychiatric implications and long-term consequences of Fetal Alcohol Spectrum Disorders. Semin Clin Neuropsych 5:177–190 (2000).
3. Sampson PD, Streissguth AP, Bookstein FL, Little RE, Clarren SK, Dehaene P, et al. Incidence of fetal alcohol syndrome and prevalence of alcohol-related neurodevelopmental disorder. Teratology. 1997;56(5):317-26.
4. Hartswood M, Jirotka M, Slack R, Voss A, Lloyd S. Working IT out in e-Science: Experiences of requirements capture in a Healthgrid project. Proceedings of Healthgrid. 2005. IOS Press, ISBM 158603510X, Page 198-209.
5. Mattes WB, Pettit SD, Sansone S, Bushel PR, Waters MD. Database Development in Toxicogenomics: Issues and Efforts. Environmental Health Perspectives. 2004; 112; 495-505.
6. Stein HD, Nadkarni P, Erdos J, Miller PL. Exploring the Degree of Concordance of Coded and Textual Data in Answering Clinical Queries from a Clinical Data Repository. Am Med Inform Assoc. 2000;7:42-54.
7. Brindis RG, Fitzgerald S, Anderson HV, Shaw RE, Weintraub WS, Williams JF. The American College of Cardiology-National Cardiovascular Data Registry (ACC-NCDR): building a national clinical data repository. J. Am. Coll. Cardiol. 2001;37;2240-2245.
8. CAS basics. 2006. http://uits.iu.edu/scripts/ose.cgi?akui.def.help.
9. The Indiana University Massive Data Storage System Service. 2006. http://storage.iu.edu/mdss.shtml.
10. High Performance Storage System. 2006. http://www.hpss-collaboration.org/hpss/index. jsp.
11. Indiana's Optical Fiber Initiative – I-Light. 2006. http://www.i-light.iupui.edu.
12. May PA, Gossage JP, White-Country M, Goodhart K, Decoteau S, Trujillo PM, Kalberg WO, Viljoen DL, Hoyme HE. Alcohol Consumption and Other Maternal Risk Factors for Fetal Alcohol Syndrome among Three Distinct Samples of Women before, during, and after Pregnancy: The Risk Is Relative. Am J Med Genet C Semin Med Genet 127C:10–20 (2004).
13. Hoyme EH, Fagerlund Á, Ervalahti N, Loimu L, Autti-Rämö I. Fetal alcohol spectrum disorders in Finland: Clinical delineation of 77 older children and adolescents and comparison with affected North American children. Presented at the Western Society for Pediatric Research (February 2005).

14. Flury-Wetherill L, Foroud T, Rogers J, Moore E. The CIFASD Consortium, Fetal Alcohol Syndrome and 3-D facial imaging: Differences among three ethnic groups. Poster presentation at 29th Annual Research Society on Alcoholism Scientific Meeting, Baltimore, MD. (Jun 23-29, 2006).
15. TeraGrid. 2006. http://teragrid.org.
16. NIH Roadmap for medical research – Overview of the NIH Roadmap. 2006. http://nihroadmap.nih.gov/overview.asp

Grid-Enabled High Throughput Virtual Screening

Nicolas Jacq[1], Vincent Breton[1], Hsin-Yen Chen[2], Li-Yung Ho[2], Martin Hofmann[3],
Hurng-Chun Lee[2], Yannick Legré[1], Simon C. Lin[2], Astrid Maaß[3],
Emmanuel Medernach[1], Ivan Merelli[4], Luciano Milanesi[4], Giulio Rastelli[5],
Matthieu Reichstadt[1], Jean Salzemann[1], Horst Schwichtenberg[4],
Mahendrakar Sridhar[3], Vinod Kasam[1,3], Ying-Ta Wu[2],
and Marc Zimmermann[3]

[1] Laboratoire de Physique Corpusculaire, IN2P3 / UMR CNRS 6533,
24 avenue des Landais, 63177 AUBIERE, France
jacq@clermont.in2p3.fr
[2] Academia Sinica,
No. 128, Sec. 2, Academic Rd., NanKang, Taipei 115, Taiwan
ywu@gate.sinica.edu.tw
[3] Fraunhofer Institute for Algorithms and Scientific Computing (SCAI),
Schloss Birlinghoven, 53754, Sankt Augustin, Germany
martin.hofmann@scai.fraunhofer.de
[4] CNR-Institute for Biomedical Technologies,
Via Fratelli Cervi 93, 20090 Segrate (Milan), Italy
luciano.milanesi@itb.cnr.it
[5] Dipartimento di Scienze Farmaceutiche, Università di Modena e Reggio Emilia,
Via Campi 183, 41100 Modena, Italy
rastelli.giulio@unimore.it

Abstract. Large scale grids for in silico drug discovery open opportunities of particular interest to neglected and emerging diseases. In 2005 and 2006, we have been able to deploy large scale virtual docking within the framework of the WISDOM initiative against malaria and avian influenza requiring about 100 years of CPU on the EGEE, Auvergrid and TWGrid infrastructures. These achievements demonstrated the relevance of large scale grids for the virtual screening by molecular docking. This also allowed evaluating the performances of the grid infrastructures and to identify specific issues raised by large scale deployment.

Keywords: large scale grids, virtual screening, malaria, avian influenza.

1 Introduction

In silico drug discovery is one of the most promising strategies to speed-up the drug development process [1]. Virtual screening is about selecting in silico the best candidate drugs acting on a given target protein [2]. Screening can be done in vitro but it is very expensive as there are now millions of chemicals that can be synthesized [3]. A reliable way of in silico screening could reduce the number of molecules required for in vitro and then in vivo testing from a few millions to a few hundreds.

W. Dubitzky et al. (Eds.): GCCB 2006, LNBI 4360, pp. 45–59, 2007.

In silico drug discovery should foster collaboration between public and private laboratories. It should also have an important societal impact by lowering the barrier to develop new drugs for rare and neglected diseases [4]. New drugs are needed for neglected diseases like malaria where parasites keep developing resistance to existing drugs or sleeping sickness for which no new drug has been developed for years. New drugs against tuberculosis are also needed as the treatment now takes several months and is therefore hard to manage in developing countries.

However, in silico virtual screening requires intensive computing, in the order of a few TFlops per day to compute 1 million docking probabilities or for the molecular modelling of 1,000 compounds on one target protein. Access to very large computing resources is therefore needed for successful high throughput virtual screening [5]. Grids now provide such resources. A grid infrastructure such as EGEE [6] today provides access to more than 30,000 computers and is particularly suited to compute docking probabilities for millions of compounds. Docking is only the first step of virtual screening since the docking output data has to be processed further [7].

After introducing the case for large scale in silico docking on grids, this paper will present the context and objectives of the two initiatives against malaria and avian influenza in chapters 2 and 3. In chapter 4, the grid infrastructures and the environments used for large scale deployment are briefly described. Results and performance of grid environment are discussed in chapter 5. In chapter 6, we will also suggest some perspectives for the coming years.

2   Large Scale in Silico Docking Against Malaria

2.1   Introduction

The number of cases and deaths from malaria increases in many parts of the world. There are about 300 to 500 million new infections, 1 to 3 million new deaths and a 1 to 4% loss of gross domestic product (at least $12 billion) annually in Africa caused by malaria. The main causes for the comeback of malaria are that the most widely used drug against malaria, chloroquine, has been rendered useless by drug resistance in much of the world [8,9] and that anopheles mosquitoes, the disease carrier, have become resistant to some of the insecticides used to control the mosquito population.

Genomics research has opened up new ways of finding novel drugs to cure malaria, vaccines to prevent malaria, insecticides to kill infectious mosquitoes and strategies to prevent development of infectious sporozoites in the mosquito [10]. These studies require more and more in silico biology; from the first steps of gene annotation via target identification to the modeling of pathways and the identification of proteins mediating the pathogenic potential of the parasite. Grid computing supports all of these steps and, moreover, can also contribute significantly to the monitoring of ground studies to control malaria and to the clinical tests in plagued areas.

A particularly computing intensive step in the drug discovery process is virtual screening which is about selecting in silico the best candidate drugs acting on a given target protein. Screening can be done in vitro using real chemical compounds, but this is a very expensive not necessarily error-free undertaking. If it could be done in silico in a reliable way, one could reduce the number of molecules requiring in vitro and then in vivo testing from a few millions to a few hundreds [11]. Advance in combinatorial chemistry has paved the way for synthesizing millions of different chemical compounds. Thus there are millions of chemical compounds available in pharmaceutical laboratories and also in a very limited number of publicly accessible databases.

The WISDOM experiment ran on the EGEE grid production service during summer 2005 to analyze one million drug candidates against plasmepsin, the aspartic protease of plasmodium responsible for the initial cleavage of human haemoglobin.

2.2 Objectives

Grid Objective

A large number of applications are already running on grid infrastructures. Even if many have passed the proof of concept level [12], only few are ready for large-scale production with experimental data. Large Hadron Collider experiments at CERN, like the ATLAS collaboration [13], have been the first to test a large data production system on grid infrastructures [14]. In a similar way, WISDOM [15] aimed at deploying a scalable, CPU consuming application generating large data flows to test the grid infrastructure, operation and services in very stressing conditions.

Docking is – along with BLAST [16] homology searches and some folding algorithms – one of the most prominent applications that have successfully been demonstrated on grid testbeds. It is typically an embarrassingly parallel application, with repetitive and independent calculations. Large resources are needed in order to test a family of targets, a significant amount of possible drug candidates and different virtual screening tools with different parameter/scoring settings. This is both a computational and data challenge problem to distribute millions of docking comparisons with millions of small compound files.

Moreover, docking is the only application for distributed computing that has prompted the uptake of grid technology in the pharmaceutical industry [17]. The WISDOM scientific results are also a means of making a demonstration of the EGEE grid computing infrastructure for the end users community, of illustrating the usefulness of a scientifically targeted Virtual Organization, and of fostering an uptake of grid technologies in this scientific area [18].

Biological Objective

Malaria is a dreadful disease caused by a protozoan parasite, plasmodium. A new strategy to fight malaria investigated within WISDOM aims at the haemoglobin metabolism, which is one of the key metabolic processes for the survival of the parasite. Plasmepsin, the aspartic protease of plasmodium, is responsible for the initial cleavage of human haemoglobin and later followed by other proteases [19]. There are ten different plasmepsins coded by ten different genes in *Plasmodium falciparum* [8]. High levels of sequence homology are observed between different plasmepsins

(65-70%). Simultaneously they share only 35% sequence homology with its nearest human aspartic protease, Cathepsin D4 [9]. This and the presence of accurate X crystallographic data make plasmepsin an ideal target for rational drug design against malaria.

Docking is a first step for in silico virtual screening. Basically, protein-ligand docking is about estimating the binding energy of a protein target to a library of potential drugs using a scoring algorithm. The target is typically a protein, which plays a pivotal role in a pathological process, e.g. the biological cycles of a given pathogen (parasite, virus, bacteria…). The goal is to identify which molecules could dock on the protein's active sites in order to inhibit its action and therefore interfere with the molecular processes essential for the pathogen. Libraries of compound 3D structures are made openly available by chemistry companies, which can produce them. Many docking software products are available either open-source or through a proprietary license.

3 In Silico Docking Against Avian Influenza

3.1 Introduction

The first large scale docking experiment focused on virtual screening for neglected diseases but new perspectives appear also for using grids to address emerging diseases. While the grid added value for neglected diseases is related to their cost effectiveness as compared to in vitro testing, grids are also extremely relevant when time becomes a critical factor. A collaboration between Asian and European laboratories has analyzed 300,000 possible drug compounds against the avian influenza virus H5N1 using the EGEE grid infrastructure in April and May 2006 [20]. The goal was to find potential compounds that can inhibit the activities of an enzyme on the surface of the influenza virus, the so-called neuraminidase, subtype N1. Using the grid to identify the most promising leads for biological tests could speed up the development process for drugs against the influenza virus.

3.2 Objectives

Grid Objective
Beside the biological goal of reducing time and cost of the initial investment on structure-based drug design, there are two grid technology objectives for this activity: one is to improve the performance of the in silico high-throughput screening environment based on what has been learnt in the previous challenge against malaria; the other is to test another environment which enables users to have efficient and interactive control on the massive molecular dockings on the grid. Therefore, two grid tools were used in parallel in this second large scale deployment. An enhanced version of WISDOM high-throughput workflow was designed to achieve the first goal and a lightweight framework called DIANE [21,22] was introduced to carry a significant fraction of the deployment for implementing and testing the new scenario.

Biological Objective

The potential for reemergence of influenza pandemics has been a great threat since the report that avian influenza A virus (H5N1) could acquire the ability to be transmitted to humans. Indeed, an increase of transmission incidents suggests a risk of human-to-human transmission [23]. In addition, the report of the development of drug resistant variants [24] is of potential concern. Two of the present drugs (oseltamivir and zanamivir) were discovered through structure-based drug design targeting influenza neuraminidase (NA), a viral enzyme that cleaves terminal sialic acid residues from glycoconjugates. The action of NA is essential for virus proliferation and infectivity; therefore, blocking its activity generates antivirus effects. To minimize non-productive trial-and-error approaches and to accelerate the discovery of novel potent inhibitors, medical chemists take advantage of modelled NA variant structures and structure-based design.

A key task in structure-based design is to model complexes of candidate compounds to structures of receptor binding sites. The computational tools for the work are based on molecular docking engines, such as AutoDock [25], to carry out a quick conformational search for small compounds in the binding sites, fast calculation of binding energies of possible binding poses, prompt selection for the probable binding modes, and precise ranking and filtering for good binders. Although docking engines can be run automatically, one should control the dynamic conformation of the macromolecular binding site (rigid or flexible) and the spectrum of the screening small organics. This process is characterized by computational and storage loads which pose a great challenge to resources that a single institute can afford.

4 The Grid Tools

4.1 The Grid Infrastructures

Compared to WISDOM which used only the EGEE infrastructure, the large scale deployment against avian influenza used three infrastructures which are sharing the same middleware (LCG2) and also common services: Auvergrid, EGEE and TWGrid.

Auvergrid

Auvergrid is regional grid deployed in the French region Auvergne. Its goal is to explore how a grid can provide the resources needed for public and private research at a regional level. With more than 800 CPUs available at 12 sites, Auvergrid hosts a variety of scientific applications from particle physics to life science, environment and chemistry.

TWGrid

TWGrid is responsible for operating a Grid Operation Centre in Asia-Pacific region. Apart from supporting the worldwide grid collaboration in high-energy physics, TWGrid is also in charge of federating and coordinating regional grid resources to promote the grid technology to the e-science activities (e.g. life science, atmospheric science, digital archive, etc.) in Asia.

EGEE

The EGEE project [6] (Enabling Grid for E-sciencE) brings together experts from over 27 countries with the common aim of building on recent advances in grid technology and developing a service grid infrastructure which is available to scientists 24 hours a day. The project aims to provide researchers in academia and industry with access to major computing resources, independent of their geographic location. The EGEE infrastructure is now a production grid with a large number of applications installed and used on the available resources. The infrastructure involves more than 180 sites spread out across Europe, America and Asia.

The two applications described in this paper were deployed within the framework of the biomedical Virtual Organization. Resource nodes available for biomedical applications scale up to 3,000 CPUs and 21 TB disk space. The resources are spread over 15 countries.

4.2 The WISDOM Production Environment

The WISDOM production environment was designed to achieve production of a large amount of data in a limited time using EGEE, Auvergrid and TWGrid middleware services. Three packages were developed in Perl and Java. Their entry points are a simple command line tool. The first package installs the application components (software, compounds database...) on the grid computing nodes. The second package tests these components. The third package monitors the submission and the execution of the WISDOM jobs.

4.3 The DIANE Framework

DIANE is a lightweight distributed framework for parallel scientific applications in a master-worker model. It assumes that a job may be split into a number of independent tasks, which is a typical case in many scientific applications. It has been successfully applied in a number of applications ranging from image rendering to data analysis in high-energy physics.

As opposed to standard message passing libraries such as MPI [26], the DIANE framework takes care of all synchronization, communication and workflow management details on behalf of the application. The execution of a job is fully controlled by the framework, which decides when and where the tasks are executed. Thus the application is very simple to program and contains only the essential code directly related to the application itself without bothering about networking details. Aiming to efficiently bridge underlying distributed computing environments and application centric user interface, DIANE itself is a thin software layer which can easily work on top of more fundamental middleware such as LSF, PBS or the grid Resource Brokers. It may also work in a standalone mode and does not require any complex underlying software.

5 Results

5.1 WISDOM Results

The WISDOM experiment ran on the EGEE grid production service from 11 July 2005 until 19 August 2005. It saw over 42 million docked compounds, the equivalent of 80 years on a single PC, in about 6 weeks. Up to 1,700 computers were simultaneously used in 15 countries around the world. FlexX [27], a commercial software with a license server, was successfully deployed on more than 1,000 machines at the same time. WISDOM demonstrated how grid computing can help drug discovery research by speeding up the whole process and reduce the cost to develop new drugs to treat diseases such as malaria.

Grid Performances
The large scale deployment was a very useful experience to identify the limitations and bottlenecks of the EGEE infrastructure. The overall grid efficiency was on average about 50%. This means that a large fraction of the jobs had to be resubmitted. This generated a significantly larger workload for the users. The different sources of failures are identified on table 1 with their corresponding rates.

In the phase where proprietary licensed software was deployed on the grid, the dominant origin of failure was the license server in charge of distributing tokens to all the jobs running on the grid. The development of a grid service to manage proprietary licensed software is under way to address this single point of failure ignored by the information system.

The second most important sources of failure were workload management and site failures, including overload, disk failure, node mis-configuration, disk space problem, air-conditioning and power cut. To improve the job submission process, an automatic resubmission of jobs was included in the WISDOM execution environment. However, the consequence of automatic resubmission was the creation of several "sinkhole" effects where all the jobs were attracted to a single node. These sinkhole effects were observed when the status of a Computing Element was not correctly described in the information system. If a Computing Element already loaded is still viewed as completely free by the Information System, it keeps receiving jobs from the Resource Broker. If the Computing Element gets down, all jobs are aborted. Even if the Computing Element can support the excessive number of jobs, the processing time is going to be very long.

The WISDOM production system developed to submit the jobs on the grid accounted for a small fraction of the failures, as did also the grid data management system. About 1 terabyte of data was produced by the 72,000 jobs submitted. Collection and registration of these output data turned out to be a heavy task. The grid data management services allowed replicating all the output files for backup. However, they did not allow storing all the results directly in a database.

Finally, unclassified failures accounted for 4% of inefficiency. This illustrates the work which is still needed to improve grid monitoring.

Table 1. Origin of failures during the WISDOM deployment with their corresponding rates

	Rate	Reasons
Success rate after checking output data	46 %	
Workload Management failure	10 %	Overload, disk failure
		Misconfiguration, disk space problem
		Air-conditioning, power cut
Data Management failure	4 %	Network / connection
		Power cut
		Other unknown causes
Sites failure	9 %	Misconfiguration, tar command, disk space
		Information system update
		Job number limitation in the waiting queue
		Air-conditioning, power cut
Unclassified	4 %	Lost jobs
		Other unknown causes
Server license failure	23 %	Server failure
		Power cut
		Server stop
WISDOM failure	4 %	Job distribution
		Human error
		Script failure

Several of the issues identified during WISDOM deployment were improved for the second large scale docking targeted on avian influenza which is described in the next chapter.

WISDOM Biological Results
Post-processing of the huge amount of data generated was a very demanding task as millions of docking scores had to be compared. At the end of the large scale docking deployment, the best 1,000 compounds based on scoring were selected thanks to post-processing ranking jobs deployed on the grid. They were inspected individually. Several strategies were employed to reduce the number of false positives. A further 100 compounds were selected after post-processing. These compounds had been selected based on the docking score, the binding mode of the compound inside the binding pocket and the interactions of the compounds to key residues of the protein.

There are several scaffolds in the 100 compounds selected after post processing. The scaffolds diphenylurea, thiourea, and guanidino analogues are most repeatedly identified in the top 1,000 compounds. Some of the compounds identified were similar to already known plasmepsin inhibitors, like the diphenylurea analogues which were already established as micro molar inhibitors for plasmepsins (Walter Reed compounds) [28]. This indicates that the overall approach is sensible and that large scale docking on computational grids has potential to identify new inhibitors. Figure 1 represents a diphenylurea analogue with a good score, well inside the binding pocket of plasmepsin, and interacting with key protein residues. The guanidino analogues can become a novel class of plasmepsin inhibitors.

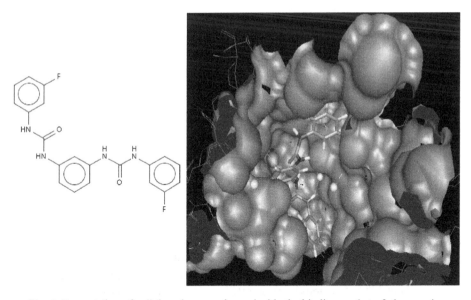

Fig. 1. Presentation of a diphenylurea analogue inside the binding pocket of plasmepsin

5.2   Results from the Large Scale Deployment Against Avian Influenza

Table 2 summarizes the achieved deployments using WISDOM and DIANE environments.

Table 2. Statistical summary of the WISDOM and DIANE activities

	WISDOM	DIANE
Total number of completed dockings	2 * 106	308,585
Estimated duration on 1 CPU	88.3 years	16.7 years
Duration of the experience	6 weeks	4 weeks
Cumulative number of grid jobs	54,000	2,585
Maximum number of concurrent CPUs	2,000	240
Number of used Computing Elements	60	36
Crunching factor	912	203
Approximated distribution efficiency	46%	84%

The WISDOM production environment was used to distribute 54,000 jobs on 60 grid Computing Elements to dock about 2 million pairs of target and chemical compound for a total amount of about 88 CPU years. Because the grid resources were used by other Virtual Organizations during the deployment, a maximum of 2,000 CPUs were concurrently running at the same time. For the DIANE part, we were able to complete 308,585 docking runs (i.e. 1/8 of the whole deployment) in 30 days using the computing resources of 36 grid nodes. A total number of 2,585 DIANE worker agents have been running as grid jobs during that period and the DIANE master concurrently maintained 240 of them. The distribution of those grid jobs in terms of the regions of the world is shown on Figure 2. About 600 gigabytes of data has been produced on the grid during the deployment.

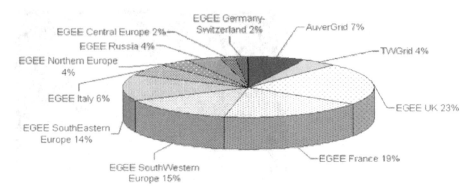

Fig. 2. Distribution of the grid jobs in different region

Grid Performances

Since a grid is a dynamic system in which the status of resources may change without central control, transient problems occur which cause job failures. In the WISDOM activity, about 83% of the jobs were reported as successfully finished according to the status logged in the grid Logging and Bookkeeping system; the observed failures were mainly due to errors at job scheduling time because of misconfiguration of grid Computing Elements. However, the success rate went down to 70% after checking the content of the data output file. The main cause for these failures was frequent last-minute errors in the transfer of results to the grid Storage Elements. Compared to the previous large scale deployment, improvement is significant as the observed success rates were respectively 77 and 63%. The last-minute error in output data transfer is particularly expensive since the results are no longer available on the grid Worker Node although they might have been successfully produced.

In DIANE, a similar job failure rate was also observed; nevertheless, the failure recovery mechanism in DIANE automated the re-submission and guaranteed a finally fully complete job. On the other hand, the feature of interactively returning part of the computing efforts during the runtime (e.g. the output of each docking) also introduces a more economical way of using the grid resources.

For the instances submitted using WISDOM production environment, the overall crunching factor was about 912. The corresponding distribution efficiency defined as the ratio between the overall crunching factor and the maximum number of concurrently running CPUs was estimated to 46%. This is due to the known issue of long job waiting time in the current EGEE production system.

The task pull model adopted by DIANE allows isolating the scheduling overhead of the grid jobs and is therefore expected to achieve a better distribution efficiency. During the deployment, DIANE was able to push the efficiency to higher than 80% within the scope of handling intermediate scale of distributed docking. A fare comparison with WISDOM can only be made if the improvement of the DIANE framework is tested in a large scale as the exercise of WISDOM.

Biological Results

Two sets of re-ranked data for each target were made (QM-MM method and selection by the study of the interactions between the compound and the target active site). The top 15% is about 45,000 compounds for each target. This set will be publicly available for the scientific community working on the avian influenza neuraminidase. The top 5% is about 2,250 compounds for each target. This set will be refined by different methods (molecular modeling, molecular dynamics…). The analysis will indicate which residue mutation is critical, which chemical fragments are preferred in the mutation sub sites and other information for lead optimization to chemists. Finally, at least 25 compounds will be assayed experimentally at the Genomic Research Center from Academia Sinica.

6 Perspectives

The results obtained by the first two high throughput virtual docking deployments have opened important perspectives. On the grid side, developments are under way to further improve the WISDOM and DIANE environments to improve the quality of service offered to the end-users. From an application point of view, beyond the necessity to analyze the results obtained on malaria and avian influenza, it is particularly relevant to further enable the deployment of virtual screening on large scale grids.

6.1 Perspective on the Grid Side: WISDOM-II

The impact of the first WISDOM deployment has significantly raised the interest of the research community on neglected diseases so that several laboratories all around the world have expressed interest to propose targets for a second large scale deployment called WISDOM-II. Contacts have been established with research groups from Italy, United Kingdom, Venezuela, South Africa and Thailand and discussions are under way to finalize the list of targets to be docked.

Similarly, several grid projects have expressed interest to contribute to the WISDOM initiative by providing computing resources (Auvergrid, EELA (http:// www.eu-eela.org/), South East Asia Grid, Swiss Biogrid (http://www.swissbiogrid. com/)) or by contributing to the development of the virtual screening pipeline (Embrace (http:// www.embracegrid.info/), BioinfoGRID (http://www.bioinfogrid.eu/)).

6.2 From Virtual Docking to Virtual Screening

While docking methods have been significantly improved in the last years by including more through compound orientation search, additional energy contributions and/or refined parameters in the force field, it is generally agreed that docking results need to be post-processed with more accurate modeling tools before biological tests are undertaken. Molecular dynamics (MD) [29] has great potential at this stage: firstly, it enables a flexible treatment of the compound/target complexes at room temperature for a given simulation time, and therefore is able to refine compound orientations by finding more stable complexes; secondly, it partially solves conformation and orientation search deficiencies which might arise from docking; thirdly, it allows the re-ranking of molecules based on more accurate scoring functions.

Figure 3 illustrates how the best hits coming out the docking step need to be further processed and analyzed with MD.

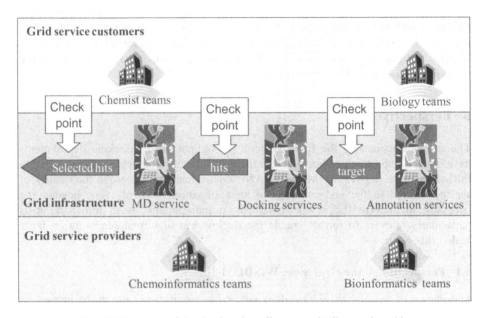

Fig. 3. First steps of the *in silico* drug discovery pipeline on the grid

However, for the same number of compounds, MD analysis requires much heavier computing than docking. Consequently, MD can only be applied to a restricted number of compounds, usually the best hits coming out of the docking step. MD and subsequent free-energy analysis most often changes significantly the scoring of the best compounds and it is therefore very important to apply it to as many compounds as possible. As a consequence, grids appear very promising to improve the virtual screening process by increasing the number of compounds that will be processed using MD.

For instance, running MD analysis with Amber 8 [30] software on one protein target and 10,000 compounds requires about 50 CPU years for a model with a few thousands atoms and a few tens of thousands of steps. This process produces only tens gigabytes of data. One run with only one compound takes 44 hours (computing time heavily depends on choice of conditions, like explicit water simulations, or generalized born simulations).

Both grids of clusters such as EGEE and grids of supercomputers such as DEISA (http://www.deisa.org/) are relevant for MD computing. MD computations of large molecules are significantly faster on a supercomputer.

Within the framework of the BioinfoGRID European project, focus will be put on the reranking of the best scoring compounds coming out of WISDOM. The goal will be to deploy on at least one of the European grid infrastructures MD software to rerank the best compounds before in vitro testing.

7 Conclusion

Large scale grids offer unprecedented perspectives for in silico drug discovery. This paper has presented pioneering activities in the field of grid enabled virtual screening against neglected and emerging diseases in Europe. This activity started with a large scale docking deployment against malaria on the EGEE infrastructure in 2005, which was followed by another large scale deployment focused on avian influenza in the spring of 2006. A second large scale deployment on neglected diseases is foreseen to take place in the fall: it will involve several European projects, which will bring additional resources and complementary contributions in order to enable a complete virtual screening process on the grid.

These deployments were achieved using two different systems for submission and monitoring of virtual docking jobs. The WISDOM system was designed to achieve production of a large amount of data in a limited time while the DIANE framework was designed as a lightweight distributed framework for parallel scientific applications in master-worker model. Both systems were able to provide high throughput virtual docking. Development of a new system merging functionalities from both WISDOM and DIANE frameworks is under way in the perspective of WISDOM-II, second large scale deployment on neglected diseases.

Acknowledgment. This work was supported in part by Auvergrid, EGEE and TWGrid projects. EGEE is funded by the European Union under contract INFSO-RI-508833. Auvergrid is funded by Conseil Régional d'Auvergne. The TWGrid is funded by the National Science Council of Taiwan. This work took place in collaboration with the Embrace network of excellence and the BioinfoGRID project. The authors express particular thanks to the site managers in EGEE, TWGrid, AuverGrid for operational supports, the LCG ARDA group for the technical support of DIANE, and the Biomedical Task Force for its participation to the WISDOM deployment. The following institutes contributed computing resources to the data challenge: ASGC (Taiwan); NGO (Singapore); IPP-BAS, IMBM-BAS and IPP-ISTF (Bulgaria); CYFRONET (Poland); ICI (Romania); CEA-DAPNIA, CGG, IN2P3-CC, IN2P3-LAL, IN2P3-LAPP and IN2P3-LPC (France); SCAI (Germany); INFN (Italy); NIKHEF, SARA and Virtual Laboratory for e-Science (Netherlands); IMPB RAS (Russia); UCY (Cyprus); AUTH FORTH-ICS and HELLASGRID (Greece); RBI (Croatia); TAU (Israel); CESGA, CIEMAT, CNB-UAM, IFCA, INTA, PIC and UPV-GryCAP (Spain); BHAM, University of Bristol, IC, Lancaster University, MANHEP, University of Oxford, RAL and University of Glasgow (United Kingdom).

References

1. BCG Estimate: A Revolution in R&D, The Impact of Genomics. (2001)
2. Lyne, P.D.: Structure-based virtual screening: an overview. Drug Discov. Today 7 (2002) 1047–1055
3. Congreve, M., et al.: Structural biology and drug discovery. Drug Discov Today 10 (2005) 895-907
4. Nwaka, S., Ridley, RG.: Virtual drug discovery and development for neglected diseases through public-private partnerships. Nat Rev Drug Discov 2 (2003) 919-28
5. Chien, A., et al.: Grid technologies empowering drug discovery. Drug Discovery Today 7 (2002) 176-180
6. Gagliardi, F., et al.: Building an infrastructure for scientific Grid computing: status and goals of the EGEE project. Philosophical Transactions: Mathematical, Physical and Engineering Sciences 363 (2005) 1729-1742 and http://www.eu-egee.org/
7. Ghosh, S., et al.: Structure-based virtual screening of chemical libraries for drug discovery. Curr Opin Chem Biol. 10 (2006) 194-202
8. Coombs, G. H., et al.: Aspartic proteases of plasmodium falciparum and other protozoa as drug targets. Trends parasitol. 17 (2001) 532-537
9. Weisner, J., et al.: Angew. New Antimalarial drugs, Chem. Int. 42 (2003) 5274-529
10. Curtis, C.F., Hoffman, S.L.: Science 290 (2000) 1508-1509
11. Spencer, R.W.: Highthroughput virtual screening of historic collections on the file size, biological targets, and file diversity. Biotechnol. Bioeng 61 (1998) 61-67
12. Jacq, N., et al.: Grid as a bioinformatics tool, Parallel Computing. 30 (2004) 1093-1107
13. Campana, S., et al.: Analysis of the ATLAS Rome Production Experience on the LHC Computing Grid. IEEE International Conference on e-Science and Grid Computing (2005)
14. Bird, I., et al.: Operating the LCG and EGEE production Grids for HEP. Proceedings of the CHEP'04 Conference (2004)
15. Breton, V., et al.: Grid added value to address malaria. Proceedings of the 6-th IEEE International Symposium on Cluster Computing and the Grid 40 (2006) and http://wisdom. healthgrid.org/.
16. Altschul, S.F., et al.: Basic local alignment search tool. J. Mol. Biol. 215 (1990) 403-410
17. Ziegler, R.: Pharma GRIDs: Key to Pharmaceutical Innovation ?, Proceedings of the HealthGrid conference 2004 (2004)
18. Jacq, N., et al.: Demonstration of In Silico Docking at a Large Scale on Grid Infrastructure. Studies in Health Technology and Informatics 120 (2006) 155-157
19. Francis, S. E., et al.: Hemoglobin metabolism in the malaria parasite plasmodium falciparum. Annu.Rev. Microbiol. 51 (1997) 97-123
20. Lee, H.-C., et al.: Grid-enabled High-throughput in silico Screening against influenza A Neuraminidase. to be published in IEEE Transaction on Nanobioscience (2006)
21. Moscicki, J.T.: DIANE - Distributed Analysis Environment for GRID-enabled Simulation and Analysis of Physics Data. NSS IEEE 2004 (2004)
22. Moscicki, J.T., et al.: Biomedical Applications on the GRID: Efficient Management of Parallel Jobs. NSS IEEE 2003 (2003)
23. Li, K.S., et al.: Genesis of a highly pathogenic and potentially pandemic H5N1 influenza virus in eastern Asia. Nature 430 (2004) 209-213
24. de Jong, M. D., et al.: Oseltamivir Resistance during Treatment of Influenza A (H5N1) Infection. N. Engl. J. Med. 353 (2005) 2667-72

25. Morris, G. M., et al.: Automated Docking Using a Lamarckian Genetic Algorithm and Empirical Binding Free Energy Function. J. Computational Chemistry 19 (1998) 1639-1662

26. Gropp, W., Lusk, E.: Dynamic process management in an MPI setting. Proceedings of the 7th IEEE Symposium on Parallel and Distributed Processing (1995)

27. Rarey, M., et al.: A fast flexible docking method using an incremental construction algorithm. J Mol Biol 261: (1996) 470-489

28. Silva, A.M., et al.: Structure and inhibition of plasmepsin II, A haemoglobin degrading enzyme from Plasmodium falciparum. Proc. Natl. Acad. Sci. USA 93 (1996) 10034-10039

29. Lamb, M. L., Jorgensen, W. L.: Computational approaches to molecular recognition. Curr. Opin. Chem. Biol. 1 449 (1997)

30. Case, D.A., et al.: The Amber biomolecular simulation programs. J. Computat. Chem. 26 (2005) 1668-1688

Grid Computing for the Estimation of Toxicity: Acute Toxicity on Fathead Minnow (*Pimephales promelas*)

Uko Maran[1], Sulev Sild[1], Paolo Mazzatorta[2], Mosé Casalegno[3], Emilio Benfenati[3], and Mathilde Romberg[4]

[1] Department of Chemistry, University of Tartu, Jakobi 2, Tartu 51014, Estonia
{uko.maran, sulev.sild}@ut.ee
[2] Chemical Food Safety Group, Dep. of Quality & Safety, Nestlé Research Center, P.O. Box 44, CH-10000 Lausanne 26, Switzerland
paolo-francesco.mazzatorta@rdls.nestle.com
[3] Istituto di Ricerche Farmacologiche "Mario Negri", Via Eritrea 62, 20157 Milano, Italy
{casalegno, benfenati}@marionegri.it
[4] School of Biomedical Sciences, University of Ulster, Cromore Road, Coleraine BT52 1SA, Northern Ireland
me.romberg@ulster.ac.uk

Abstract. The computational estimation of toxicity is time-consuming and therefore needs support for distributed, high-performance and/or grid computing. The major technology behind the estimation of toxicity is quantitative structure activity relationship modelling. It is a complex procedure involving data gathering, preparation and analysis. The current paper describes the use of grid computing in the computational estimation of toxicity and provides a comparative study on the acute toxicity of fathead minnow (*Pimephales promelas*) comparing the heuristic multi-linear regression and artificial neural network approaches for quantitative structure activity relationship models.

Keywords: modelling and prediction, chemistry, QSAR, molecular descriptors, workflow, distributed computing.

1 Introduction

Grid and high-performance computing enters many areas of computational biology and biotechnology. One such example is the estimation of toxicity of chemicals in the environment and also in organism when determining the side-effects of drugs or drug candidates. One of the key technologies behind the estimation of toxicity is quantitative structure-activity relationship (QSAR) modelling. The awareness of this technique is increasing in the scientific world and even in regulatory bodies, because it can offer alternatives to *in vitro* and *in vivo* experimental models, which are time-consuming, expensive, and subject to controversy on ethical aspects of animal testing.

W. Dubitzky et al. (Eds.): GCCB 2006, LNBI 4360, pp. 60–74, 2007.

During the last decades thousands of QSAR models have been published for various endpoints, chemical structures, descriptors, and mathematical representation. In spite of their value only few are actually distributed and used. In fact, the critical issue of QSAR models is the lack of a common and agreed protocol and the spread of competitive techniques involved, in other words the *portability* of the model [1]. This is because QSAR models dramatically depend on the 3D representation of chemical structure, on the chemical descriptor, the variable selection, and the algorithm used for the generation of the model. Furthermore models typically require several steps, using several software packages and the transforming and merging of data. Mistakes are possible, since these steps are commonly performed manually. Often different formats are requested for different software, and it may happen that results vary on the basis of the used format. These factors severely affect the accuracy and reproducibility of the results.

Although some commercial QSAR models exist for specific endpoints, there is a need for a more general, flexible tool, which should be at the same time standardised and user-friendly.

Grid computing combined with high-performance computing can provide valuable improvements, for its capability to integrate several operations provided by different modules, adding the possibility to use computers which are geographically distributed. Thus grid computing can be a viable solution for the above mentioned problems for a general reproducible QSAR strategy. In recent years grid computing has been applied to many scientific areas with different requirements, including the whole universe of life science and bioinformatics. Examples are the prediction of protein folding in the Folding@Home project [2,3], sequence analysis [4,5,6], high-throughput docking screening of around 400000 potential protein kinase CK2 inhibitors [7] and high-throughput calculation of protein-ligand binding affinities [8]. Grid computing is also used for the parameter optimisation of biochemical pathways and the allergenicity prediction of transgenic proteins [9].

Our efforts towards the grid enabling of the tools necessary for the molecular design and engineering, including QSARs, have resulted in the OpenMolGRID system [10,11]. Although the grid enabling of required tools is important, one of the challenges is the data management. Within OpenMolGRID this issue has been addressed by research towards the unified data warehouse concept [12]. Another challenge is related to making geographically distant grid applications work together in the framework of one virtual organisation. Realisation of this effort resulted in the XML based simple definition of complex scientific workflows [13,14] capable of using grid middleware functionality while assigning required grid resources automatically.

Within the current publication we use grid enabled applications linked via workflows for modelling the toxicity of chemicals using multi-linear and non-linear QSAR approaches. Our purpose is to test the performances of grid-enabled QSAR models toward the estimation of a well-known toxicological end-point, the acute toxicity on fathead minnow. This test organism is well-known in aquatic environmental risk assessment, and used as surrogate for other aquatic species [15], and many QSAR models are available in literature [16,17,18]. It therefore provides the ideal candidate for testing the application of grid-based architectures to toxicological studies. Section 2 covers the toxicity prediction process and Section 3 its requirements on grid computing and solutions covering data management and workflow issues. Section 4 provides insight into the elements of use case of toxicity estimation for the Fathead Minnow, Section 5 compares the received results with already published work and Section 6 concludes the paper.

2   Predictive Models for the Toxicity Prediction

A most widely used technique behind the predictive models for the toxicity prediction is QSAR. The QSAR methodology assumes that the biological activity (e.g. toxicity) of chemicals is determined by their molecular structures and there exists a mathematical relationship (Eq. 1) between them,

$$P = f(s), \tag{1}$$

where P is the modelled activity and s is the numerical representation (i.e. molecular descriptors) of the molecular structure. The main challenges in the development of QSAR models are the selection of appropriate molecular descriptors for the characterisation of structures and the establishment of the mathematical relationship between the activity and selected molecular descriptors. This is rather complicated task since it requires a solid understanding in various scientific disciplines and involves combined application of multiple software programs [19,20].

Toxicity prediction with QSAR methodology is a complex application consisting of two major tasks: (i) the development of QSAR models based on available experimental data and (ii) the application of these models for predicting the toxicity of new molecules. Fig. 1 shows the coarse-grain steps the processing pipelines consist of. The process starts with the design of data sets for the model development and predictions which may require access to a variety of heterogeneous data sources such as flat files, spread sheets, journal publications, and relational databases. The next steps for both tasks are the geometry optimisation of molecular structures and the calculation of molecular descriptors. Both these steps are data parallel, because each molecule is processed independently from others. This makes them ideal candidates for distributed computing. Once the experimental values and molecular descriptors are available, statistical methods are applied for the development of the actual mathematical model (Eq. 1). Finally the developed QSAR models can be used for predicting toxicity values of chemicals from their molecular descriptor values. All these steps can be performed with many different software packages which have proprietary input and output formats and configuration options, so that data format conversions are necessary to map the output of one step to the input data of the successor step in the pipeline. Currently many users manually carry out those pipelines or workflows.

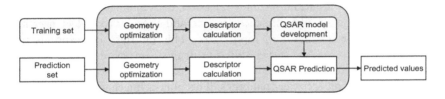

Fig. 1. Schematic representation of the workflow for the toxicity prediction

3   Grid Computing for Prediction Models

With respect to the characteristics mentioned above, the grid computing approach is an ideal solution to meet the requirements to speed-up and simplify the model building and toxicity estimation processes. Grids in general provide access to geographically distributed heterogeneous compute and data resources and allow for their coordinated use [21]. This includes tools and interfaces for accessing the different resources including software packages, scheduling and monitoring tasks, work-flow management, data transfer, security and appropriate user interfaces. Grid middleware establishes an abstraction layer between resources and the user. Therewith the user is presented with uniform interfaces to access resources, e.g. execute a structure optimisation or access a relational database. The coordinated use of resources is achieved through modelling and executing application processes as workflows.

Grid based systems for drug discovery are emerging. They differ in their focus, most of them concentrate on molecular docking, and only a few are designed to cover environmental toxicology. The Drug Discovery Grid project [22] conducts screening of chemical databases to identify compounds appropriate for a given biological receptor. Its main application within molecular modelling is docking. The user interface provides access to it via a web portal. A virtual laboratory offering drug-lead exploration, molecular docking and access to a variety of databases is going to be provided by the Australian BioGrid project [23]. A Web portal is offered to the user for setting up molecular docking experiments, monitoring and visualisation. The Wide In Silico Docking On Malaria (WISDOM) initiative for grid-enabled drug discovery against neglected and emergent diseases [24] shows how grid computing could speed-up the whole process by *in silico* docking of a huge amount of ligands in a short time. The data generated with WISDOM is going to be mined for the most promising compounds. Drug discovery workflows are dealt with in the ^{my}Grid [25] project as one of its application scenarios. It provides an *in silico* workbench with access to relevant databases and application services such as the European Molecular Biology Open Software Suite (EMBOSS, [26]).

The OpenMolGRID project [27] has been focussing on model building for toxicity prediction to support lead identification in drug development and create standard procedures for regulatory purposes. It bases on the vertically integrated grid middleware system UNICORE [28] which offers an elaborate graphical user client with workflow creation, resource selection, data management and job monitoring. Application specific interfaces can be added easily to collect user input such as parameter settings and input data. This interface corresponds to an application resource on the server side. For drug discovery and toxicity prediction these features allow to create the processing pipelines and use all those necessary application software packages. The plug-in mechanism allows for introducing another abstraction layer to integrate classes of applications which can be handled by the same application specific plug-in on the client side [11]. The application is augmented by an XML-based metadata file which describes the application and its input and output in a standardised way. This abstraction layer is also exploited for data access: A wrapper application is developed for each data source and described by metadata. The development of prediction models requires information from a variety of heterogeneous data sources which use different nomenclature and data types. Therefore a data warehouse [12] has been developed to house the cleansed data for QSAR modelling in a ready to use format. Major support for executing workflows as described in Fig. 1 is given through the MetaPlugin [13] added to the UNICORE Client. Based on a high-level XML based workflow describing only the major steps of the pipeline, it generates a detailed UNICORE workflow matching the available grid resources. For example it uses application metadata to locate target sites with required software and data resources. It also ensures that data in correct format is delivered to the target site before the execution. This is achieved by inserting additional data transfer and data conversion tasks between subsequent tasks where required. To speed up the processing, the MetaPlugin exploits distributed computing for data parallel tasks. It splits input data sets, the subsets are processed in parallel and a synchronisation step collects the output data to be transferred to the next step in the pipeline [14]. The selection of resources is normally done by the user in the UNICORE client during job preparation. The MetaPlugin automates this static scheduling process by assigning available resources with a basic round robin algorithm. Default parameter settings and defined parameter values in the workflow specification allow for completely defined pipelines which are the key to reproducible results as needed in the process of toxicity estimation.

Within the OpenMolGRID project a testbed has been set up consisting of five to ten Linux workstations located at four different sites. Each of the different application software packages have been available at least at two different compute servers. But even without exploiting parallelism the automation of the pipeline for model building itself (e.g. using only one compute system) brought significant speed-up to the process as the shown below.

4   Estimation of Acute Toxicity on Fathead Minnow

The acute toxicity is one of commonly used measures to understand whether the chemical is harmful for the living organism when entering the surrounding environment and consequently the food chain. Therefore the alternative methods for gaining knowledge on toxic behaviour of chemicals and the methods for fast and accurate estimation of acute toxicity are vital.

The applications described below build the pipelines which start with data and end with a prediction model and predicted values. These pipelines have been specified as OpenMolGRID workflows (see section 2) as schematically depicted in Fig. 1. The dark grey area indicates the applications that use grid resources. The XML definition of workflows and more detailed examples are already given in detail elsewhere [13,14,29].

The following subsections (4.1.-4.5) detail the different steps performed: the data preparation, the identification of training and prediction data set, followed by the details of molecular structure optimisation which is the basis for the descriptor calculation process and finally the two different model building procedures (multi-linear regression and artificial neural networks).

4.1   Experimental Data

The experimental data is from the U.S. EPA (Environmental Protection Agency) database and it consists of 561 industrial organic compounds of very diverse chemical structure. The data set includes chemicals with narcotic behaviour and chemicals with specific mode of action. Different modes of toxic behaviour and mechanisms of toxic action do make the dataset particularly challenging to model.

The toxicity is expressed by median lethal concentration values (LC50) for 96 hours flow-through exposures referred to juvenile stage of fathead minnow (*Pimephales promelas*) [30]. The experimental values are expressed in mol/l and in logarithmic scale.

The following procedure has been used for the distribution of the data set into training and external prediction data sets: (i) the 561 compounds are sorted based on the experimental value and (ii) every third value is moved into the prediction set. The distribution of data points results in the following: (i) the training set has 373 compounds and (ii) the prediction set 188 compounds. The data in the prediction set is only used to test the predictive power of developed models and not used during the model development phase.

4.2   Molecular Structures

The chemical structures are pre-optimised using the Merck Molecular Force Field (MMFF) [31,32] followed by conformational search via the Monte Carlo Multiple Minimum (MCMM) search method [33,34] as implemented in MacroModel^{TM}7.0 [35]. The molecular geometries corresponding to the lowest energy conformer are selected for the further calculations.

Further, the OpenMolGRID [27] system is used for the final geometry optimisation. The molecular structures are refined using eigenvectors following the geometry optimisation procedure [36] and PM3 semi-empirical parameterisation [37] in the quantum chemical program package MOPAC 7.0 [38]. The optimisation criterion, gradient norm of 0.01 kcal/Å, is specified for each structure.

4.3   Molecular Descriptors

The semi-empirical MOPAC calculations are followed by the molecular descriptor calculation. The CODESSA Pro [39] MDC module as implemented in Open-MolGRID system is used for the calculation of molecular descriptors. On average, for each structure around 600 descriptors are calculated. The calculated set of descriptors is diverse, it contains descriptors from the following categories: constitutional, topological, geometric, electrostatic, and quantum chemical [19]. In addition, octanol-water partition coefficient, $\log K_{ow}$, is included in the descriptor pool. The experimental $\log K_{ow}$ values are used when they are available and calculated with KowWin software [40] in case they are unavailable.

4.4   Forward Selection of Descriptors to Multi-linear Regression

The statistical models are developed and evaluated by means of the best multi-linear regression (BMLR) analysis [41] module of CODESSA Pro [39] as implemented in OpenMolGRID [27]. During the BMLR procedure the pool of descriptors is cleared from insignificant descriptors ($R^2 < 0.1$) and descriptors with missing values followed by the construction of the best two-parameter regression, then the best three-parameter regression, etc. based on the statistical significance and non-collinearity criteria ($R^2 < 0.6$) of the selected descriptors to the equation. In BMLR, the descriptor scales are normalised and centred automatically with the final result given in natural scales. The final model has the best representation of the activity in the given descriptor pool within the given number of parameters. The quality of the models is assessed by the squared correlation coefficient (R^2), the cross-validated (leave-one-out) squared correlation coefficient (R_{cv}^2), the square of the standard error of the estimate (s^2) and Fisher's criterion (F).

4.5   Heuristic Back-Propagation Neural Networks for QSAR

Feed-forward multi-layer neural networks [42,43] with input, hidden and output layers are used to represent non-linear QSAR models. For the development of artificial neural network (ANN) models a back-propagation algorithm with momentum term is used to train the ANN models. A subset of compounds is moved from the above described training set to the cross-validation set that is used to stop the training process in order to avoid over-training of the neural network.

The significant descriptors for the ANN models are selected by a heuristic forward selection algorithm. The descriptor selection starts with the pre-selection of molecular descriptors. If two descriptors are very highly inter-correlated then only one descriptor is selected. In addition, descriptors with insignificant variance are rejected. The descriptor selection algorithm starts by evaluating ANN models (1x1x1) with one descriptor as input. The best models are then selected for the next step, where a new descriptor is added to the input layer and the number of hidden units is increased by one. Again, the best models are selected and this stepwise procedure is repeated until the addition of new input parameters shows no significance. Since ANN models are quite likely to converge to some local minima, each model is retrained 30 times, and a model with the lowest error is selected. As the training involves a significant amount of computations, to speed up the training process, a larger learning rate is used and the number of iterations allowed is limited in the descriptor selection stage. The best models with smallest errors are selected from this procedure and are further optimised [44,45].

5   Results and Discussion

Expectedly, the neural network model (section 5.2) outperforms the multi-linear regression model (section 5.1) and provides improved prediction quality. In addition it compares well with other models for similar data sets published in the literature. Also neural network model allows better reproducibility because most of the descriptors in the model do not depend on 3D representation of the molecule.

5.1   Model Based on Multi-linear Regression

The best multi-linear regression model (Eq. 2, Fig. 2) includes six molecular descriptors for the given set of compounds used for training. The model shows statistical quality with $R^2 = 0.73$ and $s^2 = 0.53$. The leave-one-out cross-validation of the model provided a good agreement with $R^2_{cv} = 0.72$. The toxicity of 188 compounds in the external prediction set is forecasted using the proposed model. The comparison of predicted and experimental values results in $R^2 = 0.66$ and $s^2 = 0.66$.

$$\log LC_{50} = -0.56 \log K_{ow} - 4.81 P_C^{avg} + 0.17 \Delta E -$$
$$0.007\, WNSA1[PM3] - 18.58 R_C^{min} + 2.24 \frac{HASA1}{TMSA} + 0.61 \qquad (2)$$

The members of the Eq. 2 are ordered according to the t-test and revealed determinants of molecular structure that influence the toxic behaviour of chemicals. The most important descriptor is the *octanol-water partition coefficient* ($\log K_{ow}$) that determines the penetration properties through bio-membranes. The *average bond order of a C atom* (P_C^{avg}) codifies the presence and type of chemical functional groups in the molecules together with aliphatic and aromatic nature of molecules. The ΔE is an energy gap in-between the highest occupied molecular orbital and the lowest unoccupied molecular orbital and it describes the stability of chemicals. The *weighted partial negative charged surface area* (*WNSA1[PM3]*) of the molecule is the measure of the negative charge on molecular surface exposed to the environment and is an indicator of the possible site in the molecule capable of acting as potential electron donor. The *minimum 1-electron reactivity index for a C atom* (R_C^{min}) is the descriptor most difficult to justify in the given equation. Radical reactions are unlikely to occur for the given toxic endpoint. Most probably this descriptor indirectly reveals the presence of strong electron withdrawing groups in the chemicals. The last member of the equation ($HASA1/TMSA$) describes the amount of hydrogen acceptor surface area ($HASA1$) relative to total molecular surface area ($TMSA$) and justifies the importance of hydrogen-bonding interaction in describing toxicity.

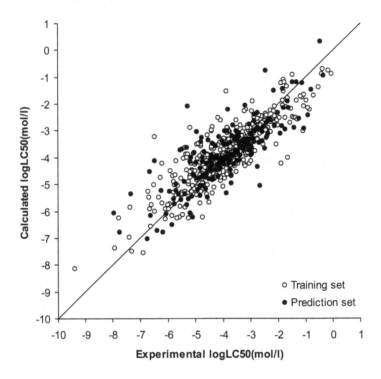

Fig. 2. Graphical representation of the multi-linear regression results

5.2   Model Based on Artificial Neural Networks

The best trade-off in-between learning and prediction quality of the neural network suggests network architecture 6x6x1 with 6 inputs, 6 hidden and 1 output layers, respectively. The statistical parameters for the training set are $R^2 = 0.78$ and $s^2 = 0.45$. The respective parameters for the external prediction set are $R^2 = 0.73$ and $s^2 = 0.53$. The overall quality of the prediction for the full data set is $R^2 = 0.76$ and $s^2 = 0.46$. The graphical representation of obtained results is given in Fig. 3.

The six molecular descriptors selected according to the heuristic procedure (section 4.5) show a good agreement with modelled toxic endpoint. Namely the *octanol-water partition coefficient* and *molecular weight* determine the penetration properties through the bio-membranes and bulk properties. The *maximum bond order of a C atom*, P_C^{max}, codifies the presence and type of chemical functional groups in the molecules along with the aliphatic and aromatic nature of the carbon backbone in molecules. This descriptor is the only one in the model that requires the 3D representation of the molecule for the calculations. The *relative number of N atoms* is a simple count of nitrogen atoms relative to the total number of atoms in the molecule. The *relative number of rings* makes a distinction between aliphatic and ring systems and also highlights their different toxic behaviour due to the planar rigid molecules and flexible molecules. Finally the *average structural information content of order 2 (ASIC2)* describes the size and branching of molecules at the second coordination sphere of the molecular graph summarising similar graph patterns.

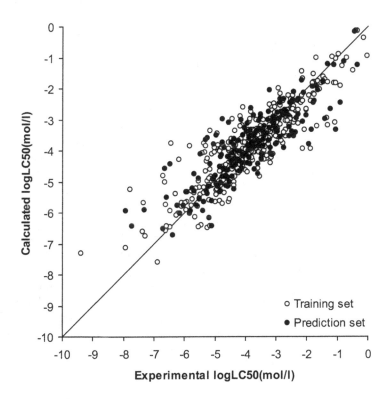

Fig. 3. Graphical representation of the results of ANN

5.3   Comparison with Previously Published Models

The comparison of results obtained in the present work with those already published for the fathead minnow dataset gives us the opportunity to measure the performances of automated, grid-enabled QSAR route with respect to traditional human-based ones. The results achieved with BMLR approach can be compared with those obtained by Netzeva et al. [46]. This study undertook multi-linear regression modelling of 568 compounds. The correlation coefficient obtained for the best four-descriptors model was $R^2 = 0.70$, while the cross-validation correlation coefficient was $R^2_{cv} = 0.70$. Despite both numbers referring to the modelling of the whole dataset, with no training/test set splitting, their magnitude is comparable to those of our BMLR model in modelling the training set ($R^2 = 0.73$, and $R^2_{cv} = 0.72$, respectively). The similar dataset of 449 compounds was also modelled by Papa et al. [47]. They used a Kohonen neural network for designing the training and prediction sets (resulting in 249 and 200 compounds in the sets, respectively) and genetic algorithms for the variable selection. The obtained squared correlation coefficient for the best five-parameter model was $R^2 = 0.81$, the cross-validation squared correlation coefficient $R^2_{cv} = 0.80$ and for the external prediction set (10 compounds) $R^2 = 0.72$. Owing to the different number of molecules and chemical composition of the sets considered, this comparison provides qualitative indications, rather than quantitative ones. Nevertheless, our purpose here is to verify whether the models generated by grid-based workflows are comparable to those generated by an expert user. To this end, we extended the comparison to ANN results, as well. In this case, for addressing the comparison we refer to the work done by Mazzatorta et al. [48] using Kohonen neural networks. The correlation coefficient obtained for the whole dataset, after the exclusion of outliers (549 molecules were considered), was $R^2 = 0.83$. This value can be compared with ours, $R^2 = 0.76$ (561 molecules, no outliers removed). It is also worth noting that the neural network used by Mazzatorta et al. contains 150 input neurons (descriptors) in comparison with 6 input neurons of the present work. Recently Vračko et al. [49] also studied a fathead minnow dataset of 551 compounds with a counter-propagation neural network. They had input of 22 neurons and compounds in the training set were predicted with $R^2 = 0.80$ and $R^2_{cv} = 0.56$. This outcome, once again, confirms that the accuracy of grid-enable models is close to that achievable by "traditional", human-based ones. In addition heuristic back-propagation neural networks show improved performance of external prediction compared to cross-validated squared correlation coefficients. One of the most valuable differences between human-based and grid-based models, however, lies in the total amount of time spent in model building. Indeed, while the model building process could take days to an expert user, it takes only few hours to the OpenMolGRID system. This makes the development of QSAR models more simple and attractive to regulators.

There is also another important advantage coming from the adoption of grid-based architectures in this field of research. As mentioned in the introduction, one of the factors hampering the wide adoption of QSAR at the regulatory level is the lack of reproducible models. Reproducibility is a necessary requirement for ensuring that QSAR will be widely accepted by regulators while facing the challenge of risk assessment. The choices made by different users are often personal and based on their own expertise. Small differences in parameter settings, or in the representation of molecular structures, could considerably affect the final results. Describing every step of the model building procedure is time-consuming and impractical. Conversely, sharing a common platform for developing models would boost cost- and time- effectiveness in QSAR development. This is one of the major strengths of the OpenMolGRID system from the point of view of toxicity prediction and legal use of QSAR models. The models we presented are not just comparable with currently published ones in terms of accuracy. They are also highly reproducible, because all steps in model building are performed in a consistent way.

6 Conclusions

Distributed, high-performance and grid computing enter many areas of computational biology and biotechnology enhancing possibilities for innovative research via integrating tools and resources from biophysics, biochemistry, biology, neuroscience, biomedicine, and environmental sciences. The driving force is the need to understand the processes in the living organism at the different levels of organisation and interaction with the surrounding and influencing environment.

In the current publication we have used distributed and grid computing in the framework of predictive toxicology for the estimation of toxicity of chemicals entering to the environment and subsequently to the living organisms. We have experienced that distributed and grid computing accelerates the process of QSAR model development. When the given task is carried out by an experienced user without grid integration, using standalone applications on single CPU with manual conversions and the transfer of data between different applications the time for the task is about five (5) days. In the case of grid integration with automated workflows the task takes on single CPU about three (3) hours. The time for the model development process can be further reduced to about half an hour by exploiting the data parallelism and including distributed computational resources e.g. using the OpenMolGRID testbed. The use of automated workflows also helps to reduce the time used for manual repetitive operations and the probability of human errors. Workflows once fixed allow the standardisation of development and application of predictive model that will facilitate the use of predictive models. Also workflows with fixed parameter settings for the applications improve the reproducibility of predictive models that allows their use for regulatory purposes.

Acknowledgements. Financial support is greatly acknowledged from the EU 5-th and 6-th framework Information Society Technologies programs (grants no. IST-2001-37238 and IST-5-033437, respectively).

References

1. Hawkins, D.M.: The Problem of Overfitting. J. Chem. Inf. Comput. Sci. 44 (2004) 1–12
2. Pande, V.S., Baker, I., Chapman, J., Elmer, S. P., Khaliq, S., Larson, S.M., Rhee, Y.M., Shirts, M.R., Snow, C.D., Sorin, E.J., Zagrovic, B.: Atomistic Protein Folding Simulations on the Submillisecond Time Scale Using Worldwide Distributed Computing. Biopolymers 68 (2003) 91–109
3. Folding@Home website at http://folding.stanford.edu/

4. Grid-BLAST website at
   http://kthgridproxy.biotech.kth.se/grid_blast/index.html
5. Chien, A., Calder, B., Elbert, S., Bhatia, K.: Entropia: Architecture and Performance of an Enterprise Desktop Grid System. J. Parallel Distrib. Comput. 63 (2003) 597–610
6. YarKhan, A., Dongarra, J.J.: Biological Sequence Alignment on the Computational Grid Using the GrADS Framework. Future Gen. Comput. Syst. 21 (2005) 980–986
7. Vangrevelinghe, E., Zimmermann, K., Schoepfer, J., Portmann, R., Fabbro, D., Furet, P.: Discovery of a Potent and Selective Protein Kinase CK2 Inhibitor by High-Throughput Docking. J. Med. Chem. 46 (2003) 2656–2662
8. Brown, S.P., Muchmore, S.M.: High-Throughput Calculation of Protein-Ligand Binding Affinities: Modification and Adaptation of the MM-PBSA Protocol to Enterprise Grid Computing. J. Chem. Inf. Model. 46 (2006) 999–1005
9. Lian, C.C., Tang, F., Issac, P., Krishnan, A.: GEL: Grid Execution Language, J. Parallel Distrib. Comput. 65 (2005) 857–869
10. Sild, S., Maran, U., Lomaka, A., Karelson M.: Open Computing Grid for Molecular Science and Engineering. J. Chem. Inf. Model. 46 (2006) 953–959
11. Mazzatorta, P., Benfenati, E., Schuller, B., Romberg, M., McCourt, D., Dubitzky, W., Sild, S., Karelson, M., Papp, A., Bágyi, I., Darvas, F.: OpenMolGRID: Molecular Science and Engineering in a Grid Context. In Proceedings of PDPTA 2004, The 2004 International Conference on Parallel and Distributed Processing Techniques and Applications, Las Vegas, USA, (2004) 775–779.
12. Dubitzky, D., McCourt, D., Galushka, M., Romberg, M., Schuller, B.: Grid-enabled Data Warehousing for Molecular Engineering. Parallel Computing 30 (2004) 1019–1035
13. Schuller, B., Romberg, M., Kirtchakova, L. Application Driven Grid Developments in the OpenMolGRID project. In Sloot, P.M.A., Hoekstra, A.G., Priol, T., Reinefeld, A., Bubak M. (eds.): Advances in Grid Computing. Lecture Notes in Computer Science, Vol. 3470. Springer-Verlag, Berlin Heidelberg New York (2005) 23–29
14. Sild, S., Maran, U., Romberg, M., Schuller, B., Benfenati, E.: OpenMolGRID: Using Automated Workflows in GRID Computing Environment. In Sloot, P.M.A., Hoekstra, A.G., Priol, T., Reinefeld, A., Bubak M. (eds.): Advances in Grid Computing. Lecture Notes in Computer Science, Vol. 3470. Springer-Verlag, Berlin Heidelberg New York (2005) 464–473
15. ECO Update, EPA, Office of emergency remedial response Hazardous Site Evaluation Division (5204G), Intermittent bulletin, Vol 2, No. 1, Publication 9345.0-051, (1994)
16. Russom, C.L., Bradbury, S.P. Broderius, S.J., Hammermeister, D.E., Drummond, R.A.: Predicting modes of toxic action from chemical structure: acute toxicity in the fathead minnow (Pimephales Promelas). Environ. Toxicol. Chem. 16 (1997) 948–967
17. Eldred D.V., Weikel C.L., Jurs P.C. Kaiser K.L.E.: Prediction of Fathead Minnow acute toxicity of organic compounds from molecular structure. Chem. Res. Toxicol., 12 (1999) 670–678
18. Nendza M., Russom, C.L.: QSAR modeling of the ERL-D Fathead Minnow Acute Toxicity Database. Xenobiotica 12 (1991) 147–170
19. Karelson, M. Molecular Descriptors in QSAR/QSPR, Wiley-Interscience (2000)
20. Maran, U., Sild, S.: QSAR Modeling of Mutagenicity on Non-congeneric Sets of Organic Compounds, In Dubitzky, W., Azuaje, F. (eds.): Artificial Intelligence Methods and Tools for Systems Biology, Springer, Dordrecht (2004) 19–36

21. Foster, I., Kesselman, C., Nick, J., Tuecke S.: The Physiology of the Grid: An Open Grid Services Architecture for Distributed Systems Integration, In F. Berman, G. Fox, A.J.G. Hey (eds.): Grid Computing: Making the Global Infrastructure a Reality, Wiley (2003) 171–198

22. Zhang, W., Zhang, J., Chang, Y., Chen, S., Du, X., Liu, F., Ma, F., Shen, J.; Drug Discovery Grid; UK e-Science All Hands Meeting 2005 http://www.allhands.org.uk/2005/proceedings/papers/578.pdf

23. Gibbins, H., Nadiminti, K., Beeson, B., Chhabra, R., Smith, B., Buyya, R.; The Australian BioGrid Portal: Empowering the Molecular Docking Research Community. Technical Report, GRIDS-TR-2005-9, Grid Computing and Distributed Systems Laboratory, University of Melbourne, Australia, June 13, 2005.

24. WISDOM website at http://public.eu-egee.org/files/battles-malaria-grid-wisdom.pdf

25. Stevens, R., McEntire, R., Goble, C., Greenwood, M., Zhao, J., Wipat, A., Li, P.: myGrid and the Drug Discovery Process. Drug Discovery Today: BIOSILICO 2 (2004) 140–148

26. Rice, P., Longden, I., Bleasby, A.: EMBOSS: The European Molecular Biology Open Software Suite. Trends Genet 16 (2000), pp. 276–277.

27. OpenMolGRID website at http://www.openmolgrid.org/

28. Erwin, D.: UNICORE - A Grid Computing Environment. Concurrency, Practice and Experience Journal, 14, 2002, pages 1395–1410

29. Maran, U., Sild, S., Kahn, I., Takkis, K.: Mining of the Chemical Information in GRID Environment. Future Gen. Comput. Syst. 23 (2007) 76–83

30. Russom, C.L., Bradbury, S.P., Broderius, S.J., Hammermeister, D.E., Drummond, R.A.: Predicting Modes of Action from Chemical Structure: Acute Toxicity in the Fathead Minnow (Pimephales Promelas). Environmental Tox. and Chem. 16 (1997) 948–957

31. Halgren, T.A.: Merck Molecular Force Field. I.–V. J. Comput. Chem. 17 (1996) 490–519, 520–552, 553–586, 587–615, 616–641

32. Halgren, T.A.: MMFF VII. Characterization of MMFF94, MMFF94s, and Other Widely Available Force Fields for Conformational Energies and for Intermolecular-Interaction Energies and Geometries. J. Comput. Chem. 20 (1999) 730–748

33. Chang, G., Guida, W.C., Still, W.C. An Internal-Coordinate Monte Carlo Method for Searching Conformational Space. J. Am. Chem. Soc. 111 (1989) 4379–4386

34. Saunders, M., Houk, K.N., Wu, Y.D., Still, W.C., Lipton, M., Chang, G., Guida, W.C.: Conformations of Cycloheptadecane. A Comparison of Methods for Conformational Searching. J. Am. Chem. Soc. 112 (1990) 1419–1427

35. MacroModel version 7.0, Interactive Molecular Modeling System, Schrödinger, Inc., Portland (2000).

36. Baker. J.: An Algorithm for the Location of Transition States. J. Comput. Chem. 7 (1986) 385–395

37. a) Stewart, J.J.P.: Optimization of Parameters for Semi-Empirical Methods I-Method. J. Comp. Chem. 10 (1989) 209–220; b) Stewart, J.J.P.: Optimization of Parameters for Semi-Empirical Methods II-Applications. J. Comp. Chem. 10 (1989) 221–264; c) Stewart, J.J.P.: Optimization of Parameters for Semi-Empirical Methods III-Extension of PM3 to Be, Mg, Zn, Ga, Ge, As, Se, Cd, In, Sn, Sb, Te, Hg, Tl, Pb, and Bi. J. Comp. Chem. 12 (1991) 320–341.

38. Stewart, J.J.P.: MOPAC: a semiempirical molecular orbital program. J. Comput. Aid. Mol. Des. 4 (1990) 1–45

39. Codessa Pro website at http://www.codessa-pro.com/

40. Meylan, W.M., Howard, P.H., Atom/Fragment Contribution Method for Estimating Octanol-Water Partition Coefficients. J. Pharm. Sci. 84 (1995) 83–92
41. Draper, N.R., Smith, H.: Applied Regression Analysis, Wiley, New York (1981)
42. Zupan, J., Gasteiger, J.: Neural Networks in Chemistry and Drug Design: An Introduction, Second edition, Wiley-VCH, Weinheim (1999)
43. Mitchell, T.M.: Machine Learning. McGraw-Hill, New York (1997)
44. Sild, S.; Karelson, M.: A General QSPR Treatment for Dielectric Constants of Organic Compounds. J. Chem. Inf. Comput. Sci. 42 (2002) 360–367
45. Maran, U., Sild, S.: QSAR modeling of genotoxicity on non-congeneric sets of organic compounds. Artif. Intell. Rev. 20 (2003) 13–38
46. Netzeva, T.I., Aptula, A.O., Benfenati, E., Cronin, M.T.D., Gini, G., Lessigiarska, I., Maran, U., Vračko, M., Schüürmann, G.: Description of the Electronic Structure of Organic Chemicals Using Semiempirical and Ab Initio Methods for Development of Toxicological QSARs. J. Chem. Inf. Model. 45 (2005) 106–114
47. Papa, E., Villa, F., Gramatica, P.: Statistically Validated QSARs, Based on Theoretical Descriptors, for Modeling Aquatic Toxicity of Organic Chemicals in Pimephales promelas (Fathead Minnow). J. Chem. Inf. Model. 45 (2005) 1256–1266
48. Mazzatorta P., Vračko M., Jezierska A, Benfenati E.: Modelling toxicity by using supervised Kohonen neural networks. J. Chem. Inf. Comput. Sci. 43 (2003) 485–492
49. Vračko, M., Bandelj, V., Barbieri, P., Benfenati, E., Chaudhry, Q., Cronin, M., Devillers, J., Gallegos, A., Gini. G., Gramatica, P., Helma, C., Mazzatorta, P., Neagu, D., Netzeva, T., Pavan, M., Patlewicz, G., Randić, M., Tsakovska, I.. Worth, A.: Validation of Counter Propagation Neural Network Models for Predictive Toxicology According to the OECD Principles: a Case Study. SAR QSAR in Environ. Res. 17 (2006) 265–284

Peer-to-Peer Experimentation in Protein Structure Prediction: An Architecture, Experiment and Initial Results

Xueping Quan[1], Chris Walton[2], Dietlind L. Gerloff[1],
Joanna L. Sharman[1], and Dave Robertson[2,*]

[1] Institute of Structural and Molecular Biology, University of Edinburgh, UK
[2] School of Informatics, University of Edinburgh, UK

Abstract. Peer-to-peer approaches offer some direct solutions to modularity and scaling properties in large scale distributed systems but their role in supporting precise experimental analysis in bioinformatics has not been explored closely in practical settings. We describe a method by which precision in experimental process can be maintained within a peer-to-peer architecture and show how this can support experiments. As an example we show how our system is used to analyse real data of relevance to the structural bioinformatics community. Comparative models of yeast protein structures from three individual resources were analysed for consistency between them. We created a new resource containing only model fragments supported by agreement between the methods. Resources of this kind provide small sets of likely accurate predictions for non-expert users and are of interest in applied bioinformatics research.

1 Introduction: Peer-to-Peer Experimentation

In Section 4 we describe novel results obtained for a specific experiment that concerns consistency in protein structure prediction. When read by itself, Section 4 is a novel piece of analysis with a result of interest to part of the bioinformatics community. The broader novelty of this paper, however, is the way in which this result is obtained and in particular the peer-to-peer architecture used to obtain it. A peer-to-peer architecture is one in which computation is distributed across processors and in which none of the processors has an overarching coordinating role - hence coordination for a given task must be achieved via communication between processors. Section 1.1 gives our perspective on scientific experimentation as a peer-to-peer activity. Section 1.2 summarises the way in which we tackle the crucial issue of maintaining the integrity of experiments in a peer-to-peer setting. With this general approach in place, we then describe (in Section 2) its use to implement the specific experiment of Section 4. Section 3 summarises the general-purpose program used to conduct our experiments, and Section 5 concludes by summarising the broader system of which this is the first part.

* The research described in this paper was funded through a grant from the European Union Sixth Framework Programme, Information Society Technologies, sponsoring the Open Knowledge Project (`www.openk.org`).

W. Dubitzky et al. (Eds.): GCCB 2006, LNBI 4360, pp. 75–98, 2007.

1.1   Scientists as Peers in a Web Community

When conducting experimental studies (or amassing information to support experimental studies) from Internet sources, each scientist (or group) may adopt a variety of roles as information providers, consumers or modifiers. Often these roles are narrowly specific, as for example the role one adopts when canvassing trusted sources for information about specific proteins and applying assessment metrics to these that are appropriate to a particular style of experimentation. In science, the roles we adopt and the specific ways in which we discharge the obligations of those roles are fundamental to establishing peer groups of "like minded" scientists in pursuit of related goals by compatible means.

The need to be precise about such obligations is strongly felt in traditional science - hence the use of rigid conventions for description of experimental method and monitoring of its execution via laboratory notebooks, enabling experiments to be monitored, replicated and re-used. Analogous structure is beginning to emerge in Internet based science. For example the structure of Web service composition in Taverna [1] provides a record of the associations between services when using these to manipulate scientific data. Like Taverna, we describe interactions as process models. Unlike Taverna, our process models are part of a system for peer-to-peer communication in which process models describing complementary roles in experimentation are shared between peers as a means of communicating and coordinating experiments.

1.2   Scientific Coordination as Peer-to-Peer Communication

Traditionally, peer-to-peer systems have not focused on the issue of maintaining the integrity of complex, flexible processes that span groups of interacting peers. Engineering solutions have polarised into those which are highly centralised (coordinating interactions through a server) versus those which rely entirely on the sophistication (and coordinated engineering) of peers to obtain reliable processes through emergent behaviours. It is, however, possible to have a distributed, de-centralised, interaction guided by a shared, mobile model of interaction. To support this we have developed a specification language, based on a process calculus, that can describe interactions between peers and (since the language is executable in the tradition of declarative programming) can be deployed to control interactions. The language is called the Lightweight Coordination Calculus (LCC) in recognition of our aim to produce the most easily applied formal language for this engineering task.

The LCC Language. Space limitations prohibit detailed discussion of LCC, its semantics or of the mechanisms used to deploy it. For these, the reader is referred to [2]. In this paper we explain enough of LCC to take us through the bioinformatics experiment of subsequent sections.

Broadly speaking, an interaction model (or, for scientists, an experimental protocol) in LCC is a set of clauses, each of which defines how a role in the interaction must be performed. Roles are described by the type of role and an identifier for the individual peer undertaking that role, which we write formally as $a(Role, Identifier)$. The definition of performance of a role is constructed using combinations of the sequence operator ('*then*') or choice operator ('*or*') to connect messages and changes of role. Messages are either outgoing to another peer in a given role ('$\Rightarrow$') or incoming from another peer in a given role ('$\Leftarrow$'). Message input/output or change of role can be governed by a constraint defined using the normal logical operators for conjunction, disjunction and negation. Notice that there is no commitment to the system of logic through which constraints are solved - on the contrary we expect different peers to operate different constraint solvers. We explain below why we take this view in the context of peer to peer experimentation.

A conversation among a group of peers can be described as a collection of dialogue sequences between peers. The speech acts conveying information between peers are performed only by sending and receiving messages. For example, suppose a dialogue allows peer $a(r1, a1)$ to broadcast a message $m1$ to peers $a(r2, a2)$ and $a(r3, a3)$ and wait for a reply from both of these. Peer $a(r2, a2)$ is expected to reply with message $m2$ to peer $a(r1, a1)$. Peer $a(r3, a3)$ is expected to do the same. Assuming that each peer operates sequentially (an assumption that is not essential to this paper but helpful for the purpose of example) the sets of possible dialogue sequences we wish to allow for the three peers in this example are as given below, where $M_o \Rightarrow A_r$ denotes a message, M_o, sent to peer A_r and $M_i \Leftarrow A_s$ denotes a message, M_i, received from peer A_s.

$$
\textbf{For } a(r1, a1) \ : \left\{
\begin{array}{l}
\langle m1 \Rightarrow a(r2, a2),\ m1 \Rightarrow a(r3, a3),\ m2 \Leftarrow a(r2, a2),\ m2 \Leftarrow a(r3, a3)\rangle, \\
\langle m1 \Rightarrow a(r2, a2),\ m1 \Rightarrow a(r3, a3),\ m2 \Leftarrow a(r3, a3),\ m2 \Leftarrow a(r2, a2)\rangle, \\
\langle m1 \Rightarrow a(r3, a3),\ m1 \Rightarrow a(r2, a2),\ m2 \Leftarrow a(r2, a2),\ m2 \Leftarrow a(r3, a3)\rangle, \\
\langle m1 \Rightarrow a(r3, a3),\ m1 \Rightarrow a(r2, a2),\ m2 \Leftarrow a(r3, a3),\ m2 \Leftarrow a(r2, a2)\rangle
\end{array}
\right\}
$$

$\textbf{For } a(r2, a2) \ : \{\langle m1 \Leftarrow a(r1, a1),\ m2 \Rightarrow a(r1, a1)\rangle\}$
$\textbf{For } a(r3, a3) \ : \{\langle m1 \Leftarrow a(r1, a1),\ m2 \Rightarrow a(r1, a1)\rangle\}$

We can specify this dialogue using a notation, similar in style to CCS, to describe the permitted message passing behaviour of each peer. Each peer, A, is defined by a term, $A :: D$, where D describes the messages it is allowed to send. D can be constructed using the operators: D_i *then* D_j (requiring D_i to be satisfied before D_j), D_i *or* D_j (requiring a choice between D_i or D_j) or D_i *par* D_j (requiring the peer to wait until D_i and D_j both are satisfied). A specification using this notation for our example is the following set of clauses:

$$
\left\{
\begin{array}{l}
a(r1, a1) :: (m1 \Rightarrow a(r2, a2) \ par \ m1 \Rightarrow a(r3, a3)) \ then \\
\qquad\qquad (m2 \Leftarrow a(r2, a2) \ par \ m2 \Leftarrow a(r3, a3)) \\
a(r2, a2) :: m1 \Leftarrow a(r1, a1) \ then \ m2 \Rightarrow a(r1, a1) \\
a(r3, a3) :: m1 \Leftarrow a(r1, a1) \ then \ m2 \Rightarrow a(r1, a1)
\end{array}
\right\}
$$

The syntax of LCC is as follows, where *Term* is a structured term and *Constant* is a constant symbol assumed to be unique when identifying each peer:

$$Framework := \{Clause, \ldots\}$$
$$Clause := Peer :: Def$$
$$Peer := a(Type, Id)$$
$$Def := Peer \mid Message \mid Def \; then \; Def \mid Def \; or \; Def \mid Def \; par \; Def$$
$$Message := Info \Leftarrow Peer \mid Info \Rightarrow Peer$$
$$Type := Term$$
$$Id := Constant$$

The above defines a space of possible dialogues but this is larger than the space we typically wish to allow in practice, since dialogues normally assume constraints on the circumstances under which messages are sent or received. These constraints are of two kinds: proaction constraints and reaction constraints. Proaction constraints define the circumstances under which a message allowed by the dialogue framework is allowed to be sent. LCC makes no assumption about the mechanism for deciding whether these constraints hold in the current state of the peer, this being a matter for the engineer of that peer. Each constraint is of the form:

$$A : (M \Rightarrow A_r) \leftarrow C_p \tag{1}$$

where A and A_r are peer descriptors (of the form $a(Type, Id)$); M is a message sent by A addressed to A_r; and C_p is the condition for sending the message (either empty or a conjunction of sub-conditions which should hold in A). In our earlier example, to constrain peer $a(r1, a1)$ to send message $m1$ to peer $a(r2, a2)$ when condition c_1 holds in $a(r1, a1)$ we would define the proaction constraint: $a(r1, a1) : (m1 \Rightarrow a(r2, a2)) \leftarrow c_1$.

Reaction constraints define what should be true in an peer following receipt of a message allowed by the dialogue framework. As for proaction constraints, LCC makes no assumption about the mechanism for ensuring that these constraints hold in the current state of the peer. Each constraint is of the form:

$$A : (M \Leftarrow A_s) \rightarrow C_r \tag{2}$$

where A and A_s are peer descriptors (of the form $a(Type, Id)$); M is a message sent by A_s and received by A; and C_r is the reaction upon receiving the message (either null or a conjunction of sub-conditions which A should be able to satisfy). In our earlier example, to constrain peer $a(r2, a2)$ to receive message $m1$ from peer $a(r1, a1)$ when condition c_2 holds in $a(r2, a2)$ we would define the proaction constraint: $a(r2, a2) : (m1 \Leftarrow a(r1, a1)) \rightarrow c_2$.

As a LCC specification is shared among a group of peers it is essential that each peer when presented with a message from that specification can retrieve the state of the dialogue relevant to it and to that message. This is done by retaining (separately) in the specification the instances of dialogue clauses used by each peer participating in the dialogue. The principle of this is similar to unfolding in the transformation of logic programs, where we can take a clause; find a sub-goal satisfiable by other clauses; then replace this with subgoal with the subgoals of that clause. A detailed description from a logic programming viewpoint appears in [2].

Using LCC as an Experiment Protocol Language. The experiment described in this paper relies on the collation of predicted structures for yeast proteins across a number of peers and comparison of the collated data across the peers to produce a tentative assessment of the predictions. The data is filtered based on these comparisons leaving behind only predictions deemed to be reliable on these grounds. Figure 1 defines a LCC specification for our example. Note that it is specific about the message sequencing and essential constraints on this type of interaction but it leaves flexible the choice and number of peers supplying data and the forms of data lookup and filtering - so the LCC specification is a model of a class of interactions, and we can ground it in specific peers and constraints at deployment time (as we show in Section 2).

Brief Comparison to Related Approaches. Since LCC is a process calculus used in a peer to peer setting, it intersects with a variety of standards and methods. We summarise below what we believe to be the three main points of contact in bioinformatics experimentation:

Targeted systems: These are systems which adopt a peer to peer architecture but where the purpose of the peer groups is focused on a specific class of experimentation. An example is the SEED system [3] which is targeted on gene annotation. This sort of system gains accessibility (for biologists) because it is specific to that domain (although of course in time SEED may generalise). LCC differs from this approach by being a generic peer to peer coordination language in which one obtains specific forms of peer to peer coordination through description of the required interaction.

Peer to peer frameworks: These are computing frameworks that allow bioinformatics programmers an easy way to establish peer to peer networks. An example is the Chinook system [4] which allows bioinformaticians familiar with Perl to build peer to peer applications. LCC also requires a form of programming (since LCC specifications have a computational behaviour) but LCC does not depend on any specific peer to peer framework. Instead, it standardises on the knowledge representation language used to represent the interaction process and its current state(s). This allows the same LCC specification to be used across multiple frameworks.

Scientific workflow systems: These provide high level visual interfaces by which scientists may describe processes connecting Web/Grid services. Examples are the Taverna [1] and Kepler [5] systems. Their role is analogous to CASE tools in traditional software engineering: they make it easier for engineers to construct executable code for a class of problems. Like CASE tools, their weakness is that outside of that class of problems an engineer must understand deeply how a specification is structured. Although it is possible to invent a visual language that can be generically used for specifying logic-based programs (see [6] for an example) there is no evidence that a generic visualisation is easier for human communities of practise to understand than the mathematical representation from which it originated. LCC therefore attempts to provide the leanest possible such language, assuming that the

A data collator, X, accepts a request for data on a yeast protein specified by its yeast-id, I, from an experimenter, E; then X changes role to being a data collector over the set, Sp, of data sources about which it has knowledge; then it reports back the set, S, of filtered data from those sources.

A data collector takes each peer identifier, P, in turn from the given set, Sp, of data sources and changes to the role of data retriever to obtain the data set, D, from P. The set Sd contains each instance of D obtained by data retrieval.

A data retriever requests data on yeast-id, I, from data source, P, and then awaits the report of the data set, D from that source.

A data source accepts a data request and replies with the the data set obtained via lookup of its database.

$a(data\_collator, X) ::$
$\quad data\_request(I) \Leftarrow a(experimenter, E)\ then$
$\quad a(data\_collector(I, Sp, Sd), X) \leftarrow yeast\_id(I)\ \wedge\ sources(Sp)\ then$
$\quad data\_report(I, S) \Rightarrow a(experimenter, E) \leftarrow filter(Sd, S)$

$a(data\_collector(I, Sp, Sd), X) ::$
$\quad (\ null \leftarrow Sp = []\ \wedge\ Sd = []\ )\ or$
$\quad \left(\begin{array}{l} a(data\_retriever(I, P, D), X) \leftarrow Sp = [P|Rp]\ \wedge\ Sd = [D|Rd]\ then \\ a(data\_collector(I, Rp, Rd), X) \end{array} \right)$

$a(data\_retriever(I, P, D), X) ::$
$\quad data\_request(I) \Rightarrow a(data\_source, P)\ then$
$\quad data\_report(I, D) \Leftarrow a(data\_source, P)$

$a(data\_source, P) ::$
$\quad data\_request(I) \Leftarrow a(data\_retriever(I, P, D), X)\ then$
$\quad data\_report(I, D) \Rightarrow a(data\_retriever(I, P, D), X) \leftarrow lookup(I, D)$

Fig. 1. Example peer-to-peer interaction specified in LCC

vast majority of scientists will be users of LCC specifications rather than designers (which requires very effective methods for sharing LCC specifications - a topic of the OpenKnowledge project described at www.openk.org). In the sense that one could build a high level visual interface for a subset of the LCC language, LCC is complementary to workflow CASE tools.

2   Experiment Implementation

The LCC specification presented in the previous section defines how the various peers will interact during the experimental process to provide the desired outcome. This specification describes how the experiment will be performed without directly identifying the peers that will be involved. To perform the actual experiment, it is necessary to supply a set of peers that match this specification, and thereby enable us to instantiate the LCC protocol. In this section, we describe how the actual experiment was performed, and outline the computational services that we constructed to accomplish this task.

From a computational point of view, our experiment is essentially a service composition task. That is, we will construct our experiment by identifying a collection of independent services, and then compose these services together dynamically to enact our experiment. In doing so, we adhere to the popular Service Oriented Architecture (SOA) paradigm, which is commonly used in Grid computing. We use our MagentA tool to perform this dynamic service composition as it was designed specifically for this purpose, as part of the OpenKnowledge project. MagentA is effectively an interpreter for LCC specifications, where the peers are defined by Web Services. We have previously demonstrated the use of MagentA to compose services in the astronomy domain [7,8]. Nonetheless, there are a number of important differences between the experiment that we perform here, and our previous astronomy experiments. Three of the key differences are summarised below:

1. Previously, we were using web services that had already been constructed for the AstroGrid project. In this case, while the necessary data is freely available on the web, there are no web services constructed to access this data. In other words, the data is accessible using HyperText Markup Language (HTML) pages intended for humans, but there are no Web Services Description Language (WSDL) interfaces and procedures to make this data available to computation entities, e.g. peers. Therefore, it was necessary to construct our own services to query and retrieve the data through form posting and screen-scraping techniques.

2. The astronomy data was all obtained from the same source and was uniform, i.e. all data was the same format and quality. Here, we are attempting to reconcile data from three independent sources. Each of these sources has derived their data using different methods, and have classified their data in different ways. To overcome these issues, it was necessary to build our services so that they can cope with missing and incomplete data, and can present the data in a uniform way. It was also necessary to design our services so that they could place quality thresholds on the data, and exclude results which did not meet these thresholds.

3. The final difference concerns the control of the underlying services and data. In the astronomy scenario, we were closely associated with the individuals who constructed the services and gathered the data. This meant that we could ask questions about the quality and distribution of the data, and obtain advice on using the services effectively. This time, we have no such close link to the data providers, and we are simply using the data that they have made publicly available. As a result, meta-information such as the quality and coverage of the data sets had to be derived experimentally.

A diagram that illustrates the main components and services in our experiment is given in Figure 2. At the left of this diagram are the three data providers for our experiment, namely: SWISS, SAM, and ModBase. These providers all make their databases available through a standard web page (i.e. HTML) interface. We note that there is an extra pre-filtering step required for the ModBase database, and we do not operate on the ModBase data set directly. This step is described later in this document.

To enable the various data sources to be used in our experiment, we have constructed a web service companion for each of the providers: a SWISS service, a SAM service, and a ModBase service. These companion services enable us to access the data through a standard web service (i.e. WSDL) interface using MagentA. These services also provide the same abstract interface to the data sources, so that we can query them in exactly the same way. This interface corresponds to $lookup(I, D)$ in Figure 1.

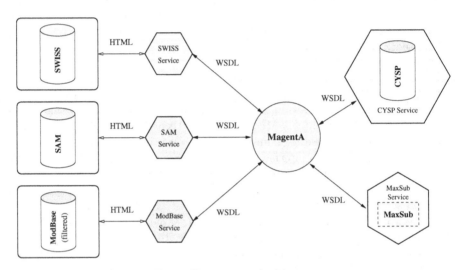

Fig. 2. Experiment Architecture

There are two additional services that we have constructed for our experiment. These services are illustrated on the right of Figure 2. The first of these is the MaxSub service, which provides a web service wrapper and WSDL interface for the MaxSub application. This application is used to perform comparisons between sequences. However, it was previously only available as a stand-alone application and could not be run over the web. The second service that we have constructed is the CYSP (Comparison of Yeast 3D Structure Predictions) service. This service is the core of our experimental process and could be viewed as corresponding to $peerX$ in Figure 1. It is responsible for querying the three data providers, invoking MaxSub to perform comparisons, and storing the results in our CYSP database. We provide our own database so that the experimental results can be reused without the need for recalculation, and for future experiment validation purposes.

3   The MagentA Architecture

Our experiment is enacted by MagentA, which is at the centre of Figure 2. As previously noted, MagentA is essentially an interpreter for LCC specifications. A specification in LCC is parameterised by peer identifiers, and in our experimental scenario we have five peers: one for each of the five key web services. During execution, these peers will interact, and as a result the services that they represent will be dynamically composed. In this section we briefly outline the architecture of MagentA, and show how dynamic composition is achieved. Further details on MagentA can be found in [7, 9, 10].

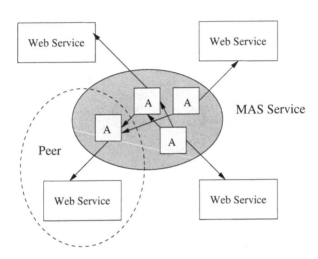

Fig. 3. Multiagent Web Services

The MagentA approach to dynamic composition is illustrated in Figure 3. MagentA is based entirely on existing Web Services technology, which permits third-party peers and services to be readily incorporated into the system without the need for modifications or plug-ins. MagentA is founded on the notion of an artificial social system, which is a technique from Multi-Agent Systems (MAS) used to coordinate large asynchronous systems of interacting entities. Peers are grouped together into societies, and communication only occurs within a specific society. The artificial society enables large systems to be constructed, and provides a controlled environment for interaction. To become a member of a society, a peer must agree to observe certain rules and conventions, and in return the peer will benefit from the other members of the society. The rules and conventions in an artificial social system are defined by explicit social norms.

A peer in MagentA is represented by a web service, and one or more proxies, which are marked **A** in our diagram above. These proxies are responsible for the communicative processes of the peers, i.e. executing the appropriate part of the LCC protocol. Each web service encapsulates the capabilities of a specific peer. At the core of the approach are MAS Services, which enable the proxies to interact. MAS services are web services that internally contain a peer communication environment. A single MAS Service is effectively responsible for a society of peers. All of the complex peer interactions happen within the MAS Service, while the external web services are accessed when necessary using standard remote procedure calls. The MAS service is also responsible for controlling entry and exit to the society, and for enforcing adherence to the social norms. In our approach, these social norms correspond directly to the LCC protocol that is executed. The MAS Service fulfils a similar role to the mediator or governor found in workflow-based composition systems, e.g. Taverna, though our use of societal concepts provides an alternative semantics.

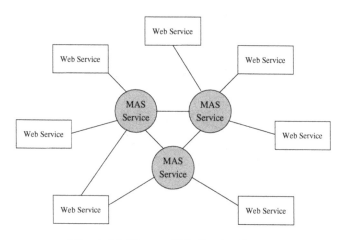

Fig. 4. Multiagent Web Service Systems

Although the MAS services impose some central control on the system, the approach is still scalable. This is because a society will typically only contain a small number of peers that directly interact, even in a large system. Figure 4 shows how large systems can be constructed from multiple MAS services. These services may define different societies, or separate instances of the same society. Furthermore, a single peer can participate in multiple societies at the same time. That is, the proxies for a web service can reside inside different MAS services. Consequently, our MagentA approach can be considered as a peer-to-peer (P2P) architecture.

As previously noted, MagentA effectively performs dynamic composition of web services. An LCC protocol is executed by a collection of proxies, acting on behalf of specific peers. As the protocol is executed, the operations inside the web services will be invoked, and the results passed to different peers. The pattern of composition is determined by the flow of control through the LCC protocol. This control flow is affected by run-time events, and so the services will not necessarily be invoked in the same way each time the protocol is executed. Thus, we have defined a protocol-based approach for dynamic service composition.

There are two main kinds of errors that can undermine dynamic service composition in MagentA. The first of these are network errors, where messages cannot be sent, or messages are not received when expected in the protocol. These kinds of errors can be handled by the timeout mechanism in LCC. Since LCC is a lightweight language, it does not directly prescribe how error recovery will be accomplished. Instead, it is up to the protocol designer to define their own recovery scheme. The timeout mechanism is the basis for the majority of such schemes (e.g. two-phase commit), and so these schemes can readily be encoded in LCC.

The second kind of error occurs when external services modify their interface definitions. This means that the service cannot be invoked, even though it is accessible on the network. If there is a choice of services with the same interface in the protocol, then the backtracking mechanism in LCC will invoke all of the alternatives in-turn. However, if there are no alternative services, then the protocol will be prematurely terminated. LCC protocols are written for specific services, and if the interface to a service changes, then the protocol must be similarly updated. We are currently investigating the use of semantic markup for services to enable such errors to be automatically handled.

The experiments described in this paper use only the basic features of MagentA and do not perform any error handling. This is because we have direct control over the web services that we use. We have constructed the five key service for our experiment: the SWISS service, the SAM service, the ModBase Service, the MaxSub service, and the CYSP service. These services provide us with a uniform way to access the data sources, and to perform computation over the data. We have also used the MagentA tools to compose these services, based around the LCC protocol that we previously presented (Figure 1). The results of the experiments are detailed in the remainder of this paper.

4   Example Experiment: Consistency-Checking in Protein Structure Prediction

Knowledge of a protein molecule's three-dimensional structure is vital for understanding its function, targeting it for drug design, etc. The two prevalent techniques for determining the 3-D coordinates of all protein atoms with high precision are X-ray crystallography and nuclear magnetic resonance spectroscopy (NMR). However, compared with the ease with which the amino acid sequences of proteins (their "1-D structures") are deduced through the sequencing of genomes and cDNA, the effort and cost required for determining a protein 3-D structure remains tremendously high. Accordingly, a number of computational biology research groups specialise in producing structural models for proteins based on their amino acid sequences alone. Where they are accurate, predicted protein structures can provide valuable clues for biological research relating to these proteins.

Thanks to regular rounds of independent assessment using newly emerging atomic protein structures as "blind" tests at the so-called CASP [11] and

CAFASP [12] experiments, protein structure prediction techniques have improved noticeably during the past decade. This is particularly true for template-directed protein structure prediction, in which the knowledge of a previously determined protein 3-D structure is used to generate a model for a different protein. If evolutionary relatedness between the two proteins can be established based on sequence similarity between a protein of interest and another protein whose 3-D structure is already known, then a model can be generated through comparative modelling. The known structure serves as a modelling template in this approach. The target protein sequence (i.e. the protein of interest) is aligned optimally onto the structural scaffold presented by the coordinate structure of this template. This typically includes the atoms making up the protein backbone, and the directionality of the side-chains (see [13] for an overview). Comparative modelling is generally considered "safe" to apply when the similarity between the sequences of target and template is sufficient to establish their alignment confidently over the whole length of the two proteins, or at least over relevant portions. However, the confidence in each individual prediction is not easily estimated at the time of the prediction. As a consequence, a substantial proportion of the models submitted for CASP/CAFASP comparative modelling targets are wrong and the degree of accuracy of their atomic coordinates varies substantially [11, 14].

In practice, biologist users of the model databases providing access to predicted protein structures are often interested in a single target protein. Such users often apply a "consistency-checking" strategy to assess whether or not to trust the predicted coordinate structures for their protein of interest by comparing the models proposed by different groups/databases to each other. Where the models agree, over the whole or a part of the protein molecule, they are deemed an approximately correct representation of the actual 3-D molecular structure of the target protein.

4.1   3-D Structural Models for Yeast Proteins

As an implemented example in which a consistency-checking experiment is undertaken we have investigated the consistency between pre-computed comparative models for the proteins encoded by the genome of the budding yeast *Saccharomyces cerevisiae*. The yeast genome sequence has been known since 1996 [15] and it is currently predicted to encode 6604 proteins. For 330 of these proteins (or fragments of them) 3-D structures for have been determined through X-ray crystallography or NMR to date. For this experiment we selected three public-domain repositories offering access to pre-computed coordinate models for yeast proteins generated by different automated methods: SWISS-MODEL [16], MODELLER (ModBase) [17], and SAM-T02/Undertaker [18]. We systematically retrieved and compared the models for all yeast proteins with to-date undetermined structures and extracted a sub-set of protein models that were "validated" by agreement between all three methods.

4.2   Data Sources

The systematic open-reading frame (ORF) names of all predicted protein-encoding genes in the yeast genome, commonly referred to as yeast-ids (YIDs), were extracted from the Saccharomyces Genome Database (SGD) [19]. The 6604 ORFs listed in SGD on 7 June 2006 were used to query the three model databases.

SWISS: The SWISS-MODEL Repository [20] is a database of annotated protein structure models generated by the SWISS-MODEL [16] comparative modelling pipeline. SWISS draws the target sequences for its entries from UNIPROT (the successor of SWISS-PROT/TrEMBL). Yeast proteins are also annotated with their YIDs. Only models for proteins of unsolved structures are accessible. If the structure of a protein was already determined experimentally (through X-ray crystallography or NMR) SWISS links directly to this structure in the PDB [21].

ModBase: This database [22] contains comparative models generated by the program ModPipe (an integration of PSI-BLAST [23] and MODELLER [17] based on protein sequences extracted from SWISS-PROT/TrEMBL. ModBase typically contains a large number of models for the same target protein that can be considered redundant and imposes only minimal quality standards. In order to streamline the procedure for this experiment and only work with models that have chances of being correct we downloaded all ModBase entries for yeast proteins and eliminated redundancy and extremely low-quality models from the set locally (see below). Note that, by contrast to SWISS, ModBase also contains models for proteins with crystallographically or spectroscopically determined structures (re-modelled onto themselves as templates and/or closely homologous proteins).

SAM: The Karplus group at UC Santa Cruz provides WWW-access to provisional models for all predicted yeast proteins using their combination of local-structure, hidden Markov model (HMM)-based fold recognition and ab initio prediction [18]. The methodology underlying fold recognition is similar to that of the comparative modelling in that a template of known structure is used. However, besides various technical differences, the application range targeted by fold recognition methods differs from that of the programs used by SWISS and ModBase in that the former specialise in predicting structures where target-template sequence similarity is too remote to detect by standard methods. Accordingly, this data source offers protein structure predictions for all ORFs of yeast (including some not listed as genes by the SGD database) and a confidence estimate for the corresponding target-template matches together with a collection of provisional (i.e. unrefined), often fragmented, coordinate models for each.

4.3   Processing/Pre-filtering of Data Sets

As described above the amount of data pertaining to each YID, and its organisation, differs quite dramatically between the three data sources. To ensure that the structural comparisons between the models could be run efficiently for the set of all available yeast models, we undertook a minimum of local processing and pre-filtering of the entries after retrieving them from the three databases. Note that most of this would be unnecessary in the more common situation where a biologist user is interested in comparing only the entries pertaining to a smaller set of YIDs, usually even only a single protein. In this case, a simple cross-check of the confidence value (which is often expressed as an E-value) associated with the model and/or a search for the best model available could be incorporated in the interaction and a larger number of pair-wise comparisons would be undertaken to extract the protein model region that is supported by the different modelling methods.

ModBase: Our MagentA interface to ModBase retrieved 3448 files with model coordinates when queried with the list of YIDs. As mentioned above these files typically include more than one 3-D model for the same protein sequence. We pre-filtered the ModBase set of models in two steps, one selecting only high-quality comparative models, and a second to eliminate redundancy. Our selection criteria for "high-quality" models were: percentage sequence identity between target and template > 20%; model score > 0.7; E-value ¡ 1E-06. The relevant values were directly accessible within the "REMARK" part of each 3-D coordinate file. Eliminating redundancy is important in cases where individual proteins are represented by multiple "high-quality" models in ModBase. As shown schematically, in Figure 5, the sequence regions covered by the different models often overlap. This can occur because different template structures were chosen (correctly, or incorrectly) each giving rise to one modelled region. To allow efficient comparison of the complete set of yeast models, we made a choice as to which one model was retained if such a redundant set was encountered. This was based on clustering of the model protein sequences extracted from the 3-D coordinate files for each YID using the program BLASTclust which is part of the BLAST suite (`www.ncbi.nlm.nih.gov/BLAST/download.shtml`). Of multiple models with > 90% pair-wise sequence identity over at least 90% (of the length of the shorter model in each comparison), only the model covering the largest sequence region was retained (Model 1 in the example in Figure 5).

By contrast to such instances of redundancy, some proteins may be represented by several models without substantial overlap. In practice this occurs often because protein structures are generally easier to determine experimentally if they only encompass a small number of structural domains. Accordingly, multi-domain protein sequences are often covered using different template structures for each domain, thus giving rise to several meaningful models. These models may not overlap at all (Figure6A), or they may overlap partly (less than 90%) with each other (Figure 6B). Multiple models of this kind were retained in our filtered ModBase model set, which contained 2546 models for 2280 yeast proteins.

SWISS: The majority of yeast proteins that can be found in SWISS database only have one associated 3-D model. By contrast to ModBase multiple entries in SWISS can be considered non-redundant, i.e. relate to multi-domain

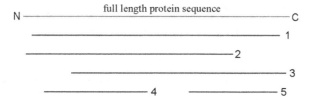

Fig. 5. Schematic showing multiple redundant models for one protein. The red line represents the full-length sequence for this protein (running from the N-terminus to the C-terminus of the molecule). Blue lines represent different models (1, 2, 3, 4, and 5) covering different, overlapping and non-overlapping, regions.

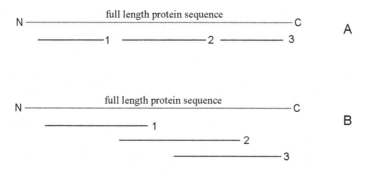

Fig. 6. Schematic showing multiple non-redundant models for one multi-domain protein. The red lines represent the full-length sequence for two multi-domain proteins. Blue lines represent models covering different regions of these sequences.

proteins. Moreover, stringent quality standards are imposed by the authors of the database. Our MagentA interface to SWISS queried the "Advanced Search" WWW-interface to the database (swissmodel.expasy.org/repository/smr.php?job=3) with the complete list of YIDs and retrieved 769 3-D models for 717 proteins when queried. In the case of multi-domain proteins, all available 3-D models were extracted. An additional 330 returns were crystallographically or spectroscopically determined structures extracted from the PDB database; these were disregarded in the structural comparisons.

SAM: Our MagentA interface to the SAM database of yeast models (www.soe.ucsc.edu/research/compbio/yeast-protein-predictions/lookup.html) returned sets of 3-D models for all 6604 YIDs. The models delivered in each set are based on different templates and ordered according to SAM-T02's confidence in the underlying target-template match. Unfortunately this organisation is not suited for extracting multiple non-redundant models from the set easily in the case of multi-domain models. For the purpose of this experiment we chose to select the top model from each set, which will usually also select the model covering the largest region of the sequence. In addition we imposed a maximum E-value cut-off of 1E-03 for the target-template match. This resulted in 2211 SAM models being considered in the structural comparisons described below.

4.4 Consistency-Checking

Pair-wise comparisons between the retained 3-D models from the three data sources relating to the same YID were carried out with the program MaxSub [24]. MaxSub performs sequence-dependent pair-wise comparisons between different 3-D structures (predicted models or known structures) of the same proteins, aiming to find the largest substructure over which the two structures superimpose well upon each other. It only considers the base (C_α) atoms of the protein side-chain and also ignores the details of the other backbone atoms. As a metric of the similarity of the two structures that are being compared, MaxSub computes a single score (referred to as Mscore below) ranging from 0 for a completely unmatched pair, to 1 for a perfect match. Since the values for Mscore are asymmetrical, i.e. dependent on which of the two proteins is considered to be the reference protein, we carried out the pair-wise comparisons in forward and reverse order. Notable deviations between the forward and reverse Mscore values arose only in cases where the lengths of the two models differed greatly and we found that the higher of the two Mscore values appeared to better reflect their degree of similarity when we inspected such cases visually. Accordingly, we always carried forward the higher value. The distance threshold parameter was set as 3.5Å throughout the analysis. For proteins whose 3-D structures were previously determined in the laboratory, there is no interest in a comparison between the X-ray/NMR structure and the 3-D models contained in ModBase and SAM, since the known structures may have been used as the modelling template. (This is different for newly determined structures which will be useful for evaluating the accuracy of the models, as is discussed below.) Pair-wise model comparisons were performed for all YIDs represented by at least one retained model in each of the three sets. In total 4556 pair-wise comparisons of the remaining yeast 3-D models returned non-zero results.

Based on the pair-wise comparisons we extracted three-way "MaxSub-supported substructures", i.e. the maximum overlap between all pair-wise matched regions for the same protein sequence. In the derivation of these substructures (illustrated schematically in Figure 7) we chose to ignore the gaps of up to 35 consecutive amino acids that were found sometimes within the regions matched in the MaxSub comparisons between two 3-D models. Such gaps were caused either by strong local deviation between the two models or missing residues in one of the models. MaxSub-supported substructures encompassing fragments of less than 45 amino acids in length were discarded to keep the number of structurally uninteresting matches (for example over only a single α-helix) as small as possible.

4.5 Results and Discussion

The detailed results of this experiment are publicly accessible via our WWW-database CYSP (Comparison of Yeast 3-D Structure Predictions, linked from www.openk.org). The records currently relate to the yeast proteins for which at least one model was retained in each of our sets of SWISS, ModBase, and SAM models after pre-filtering. Information is given regarding their associated 3-D

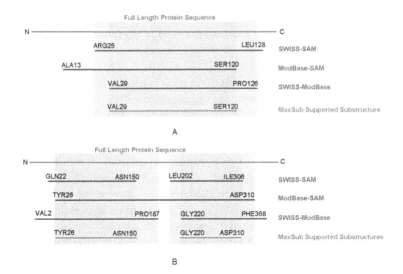

Fig. 7. Schematic illustrating the derivation of three-way MaxSub-supported substructures. Two examples are shown, a single-domain protein (A) and a multi-domain protein (B). Red lines represent full-length protein sequences. Blue lines represent the pair-wise matched regions between 3-D models for these proteins. Green lines represent the resulting MaxSub-supported substructures.

model coordinates as they can be obtained from the three repositories, which regions match pair-wise between 3-D models of the same protein by different methods according to MaxSub comparison, and the Mscores attained by the matches. For proteins where three-way agreement between the methods was found, 3-D coordinates are also provided for the model fragment spanning their Max-Sub supported substructures. To the non-computational biologist looking to find an approximate 3-D structural representation of his/her yeast protein of interest, the model fragments in this new, filtered, resource are likely to be the most relevant. While there is of course no guarantee (since there always is a chance that all three methods could have erred in the same way) they would be deemed "likely correct by consensus". This philosophy is applied widely in other areas of protein structure prediction as well, for example in secondary structure prediction, and its viability is generally supported by independently derived experimental structural information [25, 26, 27]. Attributing greater confidence to consensus predictions is certainly considered appropriate where the methods consulted are different as this was the case here.

A previous similar study by the Baker group at Washington University St. Louis compared fold predictions for yeast proteins between different fold prediction methods [28]. By contrast to our comparisons Dolinsky et al. did not carry out model superpositions but designed their SPrCY database (`agave.wustl.edu/yeast/`) for consistency checking at the template structure/fold level. Given the lower structural accuracy in general that is attained by models based on fold prediction methods (which aim primarily to identify fold resemblance to known structures in cases where no sequence similarity is detectable, and where producing a detailed model is often too difficult a problem to tackle) this is certainly justified, although it makes it impossible to directly compare our results with theirs.

We obtained 578 MaxSub-supported substructures for 545 yeast protein sequences with non-identical YIDs in this experiment. Fragments of 3-D models are most informative to the users if they span entire structural domains, or at least 3-D structurally separable parts. While some few domains are known to include less than 45 amino acids, and domain lengths spread widely, short fragments should be considered more at risk of being structurally uninformative than long fragments. As the length distribution of MaxSub-supported substructures shows (Figure 8) we would retain 136 (23%) of the corresponding model fragments even if a minimum length of 90 amino acids were imposed, rather than the 45 amino acid cut-off we chose. Thus it seems likely that the majority of the model fragments in CYSP would be useful for investigating the local structure of the yeast proteins they represent. This notion was confirmed through visual inspection of the models and is illustrated by the three examples presented in more detail below.

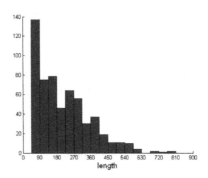

Fig. 8. Length distribution of MaxSub-supported substructures

The number of models yielding pair-wise matches is shown in Table 1. The number of matched models between SWISS and SAM is very similar to the number of three-way MaxSub-supported substructures. At first glance this coincidence may appear to reflect that only seven of the SWISS-SAM matched regions are not also supported by ModBase. However, this is not necessarily true as multi-domain proteins can give rise to different numbers of pair-wise matched regions depending on which methods are compared (as is illustrated in Figure 7B). Indeed, inspection of the results reveals that the relation between the pair-wise and three-way supported regions is not straightforward. Comparing the numbers of represented yeast proteins should be more informative and we note that the 545 proteins represented by the three-way MaxSub-supported substructure set in CYSP make up 91.8% of those represented in SWISS-ModBase; 97.5% of those in SWISS-SAM; and 91.8% of those in ModBase-SAM. While these numbers do

Table 1. Number of pair-wise matched regions between models from the three data sources (ignoring gaps). The numbers of represented yeast proteins are given in parentheses. The total number of models and proteins in each set that were considered is apparent in the diagonal. Note that comparisons were only performed for YIDs represented by at least one retained model in each filtered set.

	SWISS	ModBase	SAM
SWISS	769 (717)	649 (594)	585 (559)
ModBase		2546 (2280)	620 (594)
SAM			2211 (2211)

not differ dramatically for different method-pairs, the proportion is higher for SWISS-SAM is than for the others. If one assumed that all three-way-supported substructures are correct, but that there are likely to be more correct predictions in the set than the ones supported by all three methods, this could be explained by SAM being a slightly weaker predictor than the other two in this experiment. If this were the case it could be useful to look at the SWISS-ModBase set for additional (possibly) correct models. Alternatively, if one assumed that the three-way supported predictions are the only ones that are correct then this difference would indicate that SAM is the most useful method for preventing false models (which would make most sense if ModBase and SWISS were very similar methods). Neither of these assumptions can be expected to be entirely accurate (no known method guarantees that three-way supported predictions are actually correct) and forthcoming laboratory-determined protein structures will only provide an evaluation of a small number of the predictions in the near future. Moreover it would not be appropriate to derive more than very cautious conclusions based on this survey data. However, given the fact that the models provided by the SAM data source are deemed to be at "unrefined" stage, it is plausible that the first possibility is closer to the truth than the second. Our implementation makes it straightforward to perform a repeat experiment at a later stage of SAM-model refinement and/or to consult additional data sources in the future.

To illustrate the results accessible at CYSP we have selected three examples: YPL132W, YBR024W, and YLR132C. The 3-D models of YPL132W in SWISS, ModBase, and SAM were all generated based on the same template structure,

Table 2. Three examples of proteins included in CYSP, specified by their YIDs and SWISS-PROT identifiers, with the template structures used and the E-values the three data sources attributed to their structure predictions

YID	Protein Name	Template (E-value)		
		SWISS	ModBase	SAM
YPL132W	COX11_YEAST	1SO9A (4.5E-75)	1SO9A (3E-53)	1SO9A (4.0E-22)
YBR024	WSCO2_YEAST	2B7JB(7.9E-76)	1ON4A(4E-20)	1WPOA (2.8E-27)
YLR131	CACE2_YEAST	1NCS (4.2E-17)	1UN6B (2E-15)	2GLIA (4.7E-27)

Table 3. This table lists information extracted from the pair-wise structural comparison results by MaxSub. The missing residue column reports how many residues in Model 1 are not found in Model 2. Reported as well-matched regions are all strictly continuous sequence fragments over which Model 1 coincided well with Model 2 after 3-D structural sequence-dependent superposition.

YID	Model 1	Model 2	Missing residues	Pair-wise matched regions	MScore	MaxSub supported region
YPL132W	SWISS	SAM	2	GLU138-GLU170, ALA172-GLY219, GLU221-PHE253	0.982	GLU138-PHE253
	ModBase	SAM	2	VAL135-GLU170, ALA172-GLY219, GLU221-ALA255	0.979	
	SWISS	ModBase	0	GLU138-PHE253	0.995	
YBR024W	SWISS	SAM	44	ALA120-ALA120, GLY122-PHE125, LEU127-LYS143, SER145-TYR148, SER152-HIS153, GLU160-GLU160, LEU162-ARG164, THR166-LYS175, HIS177-ILE178, ILE180-ASP203, ILE208-SER230, TYR250-GLY263, TYR266-ARG276, GLN278-ILE279	0.660	PRO124-GLY263
	ModBase	SAM	26	PRO124-PHE125, LEU127-LYS143, SER145-TYR148, SER152-CYS154, GLU160-GLU160, LEU162-ARG164, THR166-ASP173, ILE180-ASP203, HIS205-HIS205, TYR250-GLY263	0.459	
	SWISS	ModBase	22	PRO124-LYS141, LYS143-HIS153, PRO155-PRO155, GLU160-SER170, GLN181-PHE204, PRO206-PRO206, PHE248-PRO254, GLY256-LEU262, ARG264-ARG264	0.400	
YLR131C	SWISS	SAM	0	PRO587-ILE591, VAL595-VAL595, PRO599-ASP600, LEU602-PHE614, ARG617-GLN629	0.555	(PRO606-GLN629)
	ModBase	SAM	25	LEU606-LEU606, CYS610-ASN611, PHE614-PHE614, ASN619-PHE637	0.000	
	SWISS	ModBase	10	LEU598-LEU598, CYS610-ARG616, TYR618-GLN629	0.000	

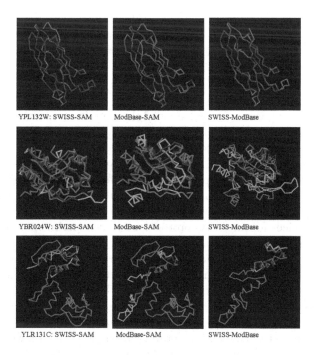

YPL132W: SWISS-SAM   ModBase-SAM   SWISS-ModBase

YBR024W: SWISS-SAM   ModBase-SAM   SWISS-ModBase

YLR131C: SWISS-SAM   ModBase-SAM   SWISS-ModBase

Fig. 9. Backbone representations of the MaxSub results for three proteins: YPL135W, YBR024WC, and YLR131C. For each pair-wise comparison between the structures of Model 1 and Model 2 (in the order MaxSub read these models), substructures in blue are the regions in Model 1 that superimpose well onto Model 2, while the other parts of the models are shown in green (Model 1) and red (Model 2).

1SO9A. By contrast, different template structures were used by each of the three model sources to model YBR024W and YLR131C (Table 2).

Pair-wise comparisons between 3-D models of YPL131W generated by SWISS, ModBase and SAM indicate that, with exception of some missing residues, the models are in perfect agreement throughout (Table 3 and Figure 9). By contrast, the three data sources only agree on the core regions of the structures predicted for YBR024W (the blue regions), and disagree otherwise (the green and red regions). Finally the three data sources disagree almost entirely on YLR131C, except over a short α-helical region (the blue regions). In this example the MaxSub-supported substructure would not be considered informative. Since it is shorter than 45 amino it was discarded and is not included in the results accessible through CYSP.

Some validation of this experiment will become possible in the future as additional structures of yeast proteins are determined in the laboratory. These newly emerging structures will allow an assessment of the accuracy of our models at least in a few cases. It will be interesting to verify, at least qualitatively, the assumption that the approach we applied extracted more accurate substructures than the models that were less well supported. To this end we are keeping track of new yeast protein structures in the PDB, and the accuracy of the corresponding models in CYSP, on www.openk.org.

5   Conclusions and Future Work

In the experiment of Section 4 we grounded the experimental protocol of Figure 1 in a specific set of services, using the MagentA system as an interpreter for the protocol. It was necessary to route all the data and analysis services through MagentA (in the way described in Figure 2) because none of the original services was equipped to interpret the protocol. We therefore incurred a small one-off cost in enabling (via WSDL and HTML) the original services to communicate with MagentA. Having done this, however, we are able to use MagentA as a proxy for the original services for any LCC protocol, so that experimenters with different ideas about how best to coordinate these (and other suitably enabled) services can implement those by altering only the protocol. Note also that the LCC protocol is separable from the mechanisms used to interpret it, and is shareable between peers during an interaction, so we can choose whether we want a single MagentA proxy for a group of services or a separate proxy for each service (giving a more or less finely grained peer-to-peer structure).

Since MagentA is capable of interpreting any LCC protocol, we can in future add to the repertoire of protocols and thus extend the capabilities of the peer-to-peer system. For example, it is straightforward to write a LCC protocol for sharing filtered data between peers. It is also straightforward to write a LCC protocol that allows queries about specific types of protein structure to be routed between peers, thus allowing networks of peers to collate, filter and propagate results. The aim of this, ultimately, is to provide a peer network that, through sharing, can produce more confident predictions faster by sharing the analyses performed earlier by others. Opportunities for applying consistency-checking, and data sharing, strategies are found in many areas of bioinformatics. The single experiment in this paper shows the immediate benefit of this on a small scale for a specific form of analysis but farther-reaching development is required to experiment effectively with large peer groups, where trust and provenance are (among other issues) important to the coherence of peer groups. This, although outside the scope of the current paper, is one of the central themes of the OpenKnowledge project (www.openk.org).

References

1. T. Oinn, M. Addis, J. Ferris, D. Marvin, M. Senger, M. Greenwood, T. Carver, K. Glover, M. Pocock, A. Wipat, and P. Li. Taverna: a tool for the composition and enactment of bioinformatics workflows. *Bioinformatics*, 20(17):3045–3054, 2004.
2. D. Robertson. Multi-agent coordination as distributed logic programming. In *International Conference on Logic Programming*, Sant-Malo, France, 2004.
3. R. Overbeek, T. Disz, and R. Stevens. The seed: a peer-to-peer environment for genome annotation. *Communications of the ACM*, 47(11):46–51, 2004.

4. S. Montgomery, J. Guan, T. Fu, K. Lin, and S. Jones. An application of peer-to-peer technology to the discovery, use and assessment of bioinformatics programs. *Nature Methods*, 2(8), 2005.
5. I. Altintas, A. Birnbaum, K. Baldridge, W. Sudholt, M. Miller, C. Amoreira, Y. Potier, and B. Ludaescher. A framework for the design and reuse of grid workflows. In P. Herrero, M. Perez, and V. Robles, editors, *International Workshop on Scientific Aspects of Grid Computing*, pages 120–133. Springer-Verlag, Lecture Notes in Computer Science 3458, 2005.
6. J. Agusti, J. Puigsegur, and D. Robertson. A visual syntax for logic and logic programming. *Journal of Visual Languages and Computing*, 9:399–427, 1998.
7. C. Walton and A. Barker. An Agent-based e-Science Experiment Builder. In *Proceedings of the 1st International Workshop on Semantic Intelligent Middleware for the Web and the Grid*, Valencia, Spain, August 2004.
8. A. Barker and R. Mann. Integration of MultiAgent Systems to AstroGrid. In *Proceedings of the Astronomical Data Analysis Software and Systems XV*, European Space Astronomy Centre, San Lorenzo de El Escorial, Spain, October 2005.
9. C. Walton. Typed Protocols for Peer-to-Peer Service Composition. In *Proceedings of the 2nd International Workshop on Peer to Peer Knowledge Management (P2PKM 2005)*, San-Diego, USA, July 2005.
10. C. Walton. Protocols for Web Service Invocation. In *Proceedings of the AAAI Fall Symposium on Agents and the Semantic Web (ASW05)*, Virginia, USA, November 2005.
11. J. Moult. A Decade of CASP: Progress, Bottlenecks and Prognosis in Protein Structure Prediction. *Curr Opin Struct Biol*, 15(3):285–289, 2005.
12. D. Fischer, L. Rychlewski, R.L. Dunbrack, A.R. Ortiz, and A. Elofsson. CAFASP3: the Third Critical Assessment of Fully Automated Structure Prediction Methods. *Proteins*, 53(6):503–516, 2003.
13. E. Krieger, S.B. Nabuurs, and G. Vriend. Homology Modeling. *Methods Biochem Anal*, 44:509–523, 2003.
14. M. Tress, I. Ezkurdia, O. Grana, G. Lopez, and A. Valencia. Assessment of Predictions Submitted for the CASP6 Comparative Modeling Category. *Proteins*, 61(7):27–45, 2005.
15. A. Goffeau, B.G. Barrell, H. Bussey, R.W. Davis, B. Dujon, H. Feldmann, F. Galibert, J.D. Hoheisel, C. Jacq, and M. Johnston. Life with 6000 Genes. *Science*, 274(5287):563–547, 1996.
16. T. Schwede, J. Kopp, N. Guex, and M.C. Peitsch. SWISS-MODEL: An Automated Protein Homology-modeling Server. *Nucleic Acids Res*, 31(13):3381–3385, 2003.
17. A. Sali and T.L. Blundell. Comparative Protein Modelling by Satisfaction of Spatial Restraints. *J Mol Biol*, 234(3):779–815, 1993.
18. K. Karplus, R. Karchin, J. Draper, J. Casper, Y. Mandel-Gutfreund, M. Diekhans, and R. Hughey. Combining Local-structure, Fold-recognition, and New Fold Methods for Protein Structure Prediction. *Proteins*, 53(6):491–496, 2003.
19. S. Weng, Q. Dong, R. Balakrishnan, K. Christie, M. Costanzo, K. Dolinski, S.S. Dwight, S. Engel, D.G. Fisk DG, and E. Hong. Saccharomyces Genome Database (SGD) Provides Biochemical and Structural Information for Budding Yeast Proteins. *Nucleic Acids Res*, 31(1):216–218, 2003.
20. J. Kopp and T. Schwede. The SWISS-MODEL Repository: New Features and Functionalities. *Nucleic Acids Res*, 34:315–318, 2006.
21. H.M. Berman, J. Westbrook, Z. Feng, G. Gilliland, T.N. Bhat, H. Weissig, I.N. Shindyalov, and P.E. Bourne. The Protein Data Bank. *Nucleic Acids Research*, 28(1):235–242, 2000.

22. U. Pieper, N. Eswar, H. Braberg, M.S. Madhusudhan, F.P. Davis, A.C. Stuart, N. Mirkovic, A. Rossi, M.A. Marti-Renom, and A. Fiser. MODBASE, a Database of Annotated Comparative Protein Structure Models, and Associated Resources. *Nucleic Acids Res*, 32:217–222, 2004.

23. S.F. Altschul, T.L. Madden, A.A. Schaffer, J. Zhang, Z. Zhang, W. Miller, and D.J. Lipman. Gapped BLAST and PSI-BLAST: a New Generation of Protein Database Search Programs. *Nucleic Acids Research*, 25(17):3389–3402, 1997.

24. N. Siew, A. Elofsson, L. Rychlewski, and D. Fischer. MaxSub: an Automated Measure for the Assessment of Protein Structure Prediction Quality. *Bioinformatics*, 16(9):776–785, 2000.

25. D. Fischer. Servers for Protein Structure Prediction. *Curr Opin Struct Biol*, 16(2):178–182, 2006.

26. R.L. Dunbrack. Sequence Comparison and Protein Structure Prediction. *Curr Opin Struct Biol*, 16(3):374–384, 2006.

27. J. Heringa. Computational Methods for Protein Secondary Structure Prediction Using Multiple Sequence Alignments. *Curr Protein Pept Sci*, 1(3):271–301, 2000.

28. T.J. Dolinsky, P.M. Burgers, K. Karplus, and N.A. Baker. SPrCY: Comparison of Structural Predictions in the Saccharomyces cerevisiae Genome. *Bioinformatics*, 20(14):2312–2314, 2004.

Gene Prediction in Metagenomic Libraries Using the Self Organising Map and High Performance Computing Techniques

Nigel McCoy[1], Shaun Mahony[2], and Aaron Golden[3]

[1] National University of Ireland, Galway, Ireland
N.McCoy1@nuigalway.ie
[2] Dept of Computational Biology, University of Pittsburgh, Pittsburgh, PA, USA
shaun.mahony@ccbb.pitt.edu
[3] National University of Ireland, Galway, Ireland
agolden@it.nuigalway.ie

Abstract. This paper describes a novel approach for annotating metagenomic libraries obtained from environmental samples utilising the self organising map (SOM) neural network formalism. A parallel implementation of the SOM is presented and its particular usefulness in metagenomic annotation highlighted. The benefits of the parallel algorithm and performance increases are explained, the latest results from annotation on an artificially generated metagenomic library presented and the viability of this approach for implementation on existing metagenomic libraries is assessed.

Keywords: Self Organising Map, Metagenomics, HPC, MPI.

1   Introduction

Metagenomics combines the technology developed to sequence individual genomes with the analysis of large scale assemblages of micro organisms taken from specific ecological environments. Traditional methods used to culture specific micro organisms are difficult as one cannot perfectly replicate a specific ecological niche within which such microbes thrive. It is estimated that up to 99 percent of the world's bacteria cannot be cultured [1][2] and so direct analysis of sequence taken from different biological niches is required. The prevalent sequencing technology is whole genome shotgun (WGS) sequencing [3] which does not require cultivation of each individual species present in an environmental sample. The WGS sequencing process yields sequence reads which in the case of metagenomic sequencing are typically 600 - 800 base pairs long, and which can be examined to identify species both known and as yet unknown, providing a 'snapshot' of evolutionary processes. Analysis also holds forth the prospect of characterising the DNA sequences of new microbial species and holds the promise of new gene discovery, with great potential implications for biotechnology.

Current techniques for annotation are based on hidden markov models (HMM) and BLAST searches. Both of these techniques use only one gene model for

W. Dubitzky et al. (Eds.): GCCB 2006, LNBI 4360, pp. 99–109, 2007.

gene annotation [4] and the methods require re-assembly of the metagenomic reads in an attempt to assemble complete genomes before annotating each of the individual assemblies. Due to the excessive number of sequence reads obtained from WGS (e.g. 10 TB per gramme sample of soil [5]) the subsequent 'mining' of the resulting sequence library presents a number of problems computationally, not least of which is dealing with the sheer volume of data. Assuming one can assemble the sequence scaffolds correctly, there is then the profoundly difficult issue of identifying distinct genomes within the meta-sequence and examining the genomes to characterise their functional genetics.

Annotation using the SOM, which has proven successful in the annotation of single genomes by characterising synonymous codon usage associated with genes within specific genomes [6] is described here. One approach is to segment the sequence reads into species specific clusters using the SOM. Recent research in [7] has shown that individual genomes display distinct 'genome signatures' in terms of the oligonucleotide frequencies occurring in the genomic sequence. Such genome signatures can be used to distinguish one microbial genome from another. Segmenting makes reassembly and hence annotation more practical. Another approach explored here, is to use the SOM to analyse the coding potential of each of the metagenomic reads directly. The proposed method skips assembly and uses multiple gene models to assess the coding potential of each of the individual shotgun sequence sized reads obtained in the shotgun sequencing process.

2   Rationale for the SOM

The task of sequencing environmental samples is complicated by the sheer volume of data retrieved and the genomic diversity present in the environmental samples. The number of times that any base is sequenced can be assumed to be Poisson distributed in accordance with the Lander-Waterman model [4]. If P is the probability of Y events occurring and λ is the average number of events (average number of bases sequenced) then the average number of times that any base is sampled y times can be given as

$$P(Y = y) = \frac{\lambda^y \times e^{-\lambda}}{y!}. \tag{1}$$

Then

$$C = \frac{L \times N}{G}, \tag{2}$$

where C is defined as the coverage (average number of times each base is sequenced) and is equivalent to λ in Equation 1, L is the length of each sequence read, N is the number of sequence reads and G is the total length of the genome or metagenome. The percentage of the genome that is not sequenced at a particular coverage C is then given by

$$P(Y = C) = \frac{C^0 \times e^{-C}}{0!} = e^{-C}. \tag{3}$$

Even at 3× coverage 5% of the target genome is not sequenced. These figures are substantial even in consideration of a single genome and the resulting gaps pose difficulties for re-assembling of the reads and sequencing of the genome. In the case of environmental samples where multiple genomes are present with perhaps several very dominant species then sampling at excessively high levels of coverage is necessary to obtain sufficient sequence for assembly of rare members. Samples may require 100-1000 fold coverage yielding libraries 500GB in size per ml of seawater and up to 10TB in size per gramme sample of soil [5].

Currently, analysis usually involves re-assembling of assemblages and annotation using HMM or BLAST techniques. The vast number of reads and the potentially large pattern variety present in the sequence data poses difficulties for these techniques [4]. In [6], it is shown that SOMs can be used for the discovery of genes within a single genome. This paper explains how the SOM can also be used to characterise the coding potential of each of the reads in the metagenome without the need to re-assemble the reads into their original genomes, hence overcoming one of the great computational challenges in metagenomic annotation. The SOM identifies genes according to multiple gene models represented in the map.

This multiple gene model approach is another advantage that distinguishes the SOM as being particularly useful for identification of genes in datasets with various codon usage patterns many of which will be previously uncharacterised and gives the SOM advantages over HMM and BLAST based methods, the disadvantages of which are highlighted in [4]. HMM methods are not suitable for use on short reads, with high gene density as is the case with metagenomic reads and also require training using closely related species. It may not be possible to determine suitable species for training and BLAST techniques will suffer from a lack of available comparable databases.

3   The SOM

The SOM, proposed by Kohonen and detailed in [8] is a commonly implemented artificial neural network (ANN) often used in pattern classification. Multi-dimensional input vectors are mapped onto a two dimensional output layer allowing for visualisation of similarities and clustering and classification of the input data. SOMs are trained to contain models that may be representative of the data in the input dataset. Each input in the dataset is characterised by a 'descriptor' which is a multi-element vector. Firstly, all the nodes on the output lattice (usually a rectangular or square grid) are initialised with vectors of random values. Training then occurs in two distinct phases - firstly an "ordering phase" performed for a proportion of the total number of user defined epochs followed by a "tuning/convergence phase". An epoch involves:

- Comparing each input in the training set (represented by the descriptor) to each node on the output lattice to find the one most similar to the input - various metrics of similarity can be used such as euclidean and cosine distance.

- Updating the 'winning' node and all nodes in a defined neighbourhood according to

$$W_{new} = W_{old} + \alpha(X_i \times W_{old}), \tag{4}$$

where W is the vector or model contained in each lattice node, α is the learning rate and X_i is the input vector. After each epoch:

- The neighbourhood size is linearly decreased (this occurs only for a defined number of epochs after which time it is set to a minimum value.)
- The learning rate (α) is decreased - the decrement size depends on the training stage. The step size decrements are large for the ordering phase and set to a relatively small value for the convergence phase.

3.1   Descriptor

The gene models contained in the SOM are based on relative synonymous codon usage (RSCU) values. For each amino acid appearing in a gene the relative usage of each synonymous codon for that amino acid is determined. The value is given by

$$RSCU_i = \frac{Obs_i}{Exp_i}, \tag{5}$$

where Obs_i is the observed number of occurrences of a particular codon, i, and Exp_i is the expected number of occurrences of that codon given the number of occurrences of that codon's corresponding amino acid and assuming that each synonym is equally likely to occur. Using the universal genetic code and disregarding the synonyms for the start and stop codons leaves 59 possible RSCU values. Each gene model or vector W in the SOM is represented by a 59 element vector made up of these values. Each of the inputs is classified in a similar manner with several modifications described in the next section.

3.2   Analysis

After training, the SOM can be used to analyse a dataset, to either cluster the similar inputs in the data or to classify the data based on the models in the map. This paper focuses on the ability of the SOM to distinguish coding sequences from non coding ones and it is this feature rather than the clustering capability of the SOM that is detailed here. Before analysis of an input dataset a set of randomly generated reads is examined using the trained SOM. Each vector of RSCU values (representative of a random 'gene') present in the random dataset is compared to each node in the map to identify the closest match in accordance with a defined metric. The cosine distance is the metric currently in use. The mean (μ) and standard deviation (σ) of the distances is determined. When analysing the input data each vector is again compared to each node in the map to determine the closest map node. The z-score for each input vector is determined according to

$$z_{score} = \frac{d - \mu}{\sigma}, \tag{6}$$

where d is the cosine distance of the winning node from the input vector, μ is the mean of the cosine distances of the random inputs from their corresponding winning nodes and σ is the standard deviation of these distances. The corresponding p value of this z-score can be found in accordance with

$$p = \frac{1}{2\pi} \times \int_{Z}^{\inf} e^{\frac{-x^2}{2}} dx, \tag{7}$$

where Z is the z-score of the input under test. This p value gives the probability that any input could result in a z-score equal to or higher than the z-score of the input under test, due to chance, given that the null hypothesis is true. In the case of this analysis the null hypothesis is that the input is non coding i.e. it is a random pattern that does not closely resemble any of the gene models in the map. The smaller the p value the more likely it is that null hypothesis is false i.e. the input is not a random pattern and is coding. The probability of the input being non random is given by

$$prob = (0.5 + \frac{1}{2\pi}) \times \int_{0}^{Z} e^{\frac{-x^2}{2}} dx. \tag{8}$$

Analysis of the inputs is complicated due to the possibility of frame shifts. To account for frame shifts each input is split into six read frames with each frame then being analysed. The samples in each frame are analysed within a sliding window of user defined length. The distance of each 'slide' of the window is also user defined. This process enables the prediction of small genes. Predictions are recorded if they have a probability greater than .1. Predictions from each of the read frames are merged as detailed in [6].

4   Parallel Approach

To facilitate annotation on realistically sized metagenomes high performance computing (HPC) techniques are utilised and a parallelised SOM based annotation tool, MetaNet, is implemented using the message passing interface (MPI). A parallelised implementation enables training of large SOMs with large numbers of homologously derived genes from various species. Ultimately SOMs trained with sufficient numbers of genes to be representative of all the genes in GenBank will be used to annotate real environmental samples such as the Sargasso sea dataset.

The parallel implementation was identified as having two distinct phases each with different architectural requirements. The training phase requires map segmentation and heavy inter process communication for updates of the correct map nodes. For SOM training a version of the algorithm from [9] is applied to metagenomic libraries to test its usefulness in metagenomic annotation. The implementation of this phase of the algorithm is more suited for symmetric multi processor (SMP) machines and other shared memory architectures. Implementation of the training phase through MPI allows for message passing through shared memory segments giving the required performance and caters for scalability, allowing for expansion of the implementation onto clusters of SMPs. The data analysis and annotation phase is a data segmentation problem that was identified as being suitable for implementation on cluster platforms. The input data (metagenome) is equally divided among each processor. The analysis phase is adapted to allow analysis of the gene content of individual sequence reads as opposed to entire genomes as before. The parallel implementation makes viable the training of large SOMs and makes the annotation of very large datasets, such as metagenomic libraries, possible.

4.1   MPI Implementation

MPI is a commonly used interface providing functions for message passing that allow for code parallelisation based on interprocess communication. The interface is implementable on clusters, single SMP machines and clusters of SMPs. An outline of the features of MPI is given in [10]. Both the training and data analysis/annotation stages of MetaNet have been adapted for parallel operation using MPI. The annotation phase requires that each process contain the entire trained SOM. The input data is shared out between each process and each segment annotated without the need to communicate with other processors. This lack of inter process communication means that this stage can be carried out on a cluster platform which may be more readily available than machines using an SMP architecture.

Training the SOM using the method in [9] requires relatively intense inter process communication. Each processing node contains a segment of the SOM represented as a one dimensional array. During training, the input vectors are read in from a disk which allows for parallel read access. Each processing node determines the lattice node that most closely resembles the input vector as described in section 3. The index and distance from the input vector of each of the 'local winners' are sent to a master node which determines which of the local winners has the minimum distance from the input vector. This is determined to be the 'global winner'. It's i and j coordinates are determined in the context of the map in its entirety. The coordinates are broadcast to all processing nodes, each of which determines if any of the lattice nodes falling within the neighbourhood surrounding the winner "reside" on that processing node and updates these nodes according to Eq. 4. This process is illustrated in Fig 1.

A HPC approach is essential because the metagenomic problem dictates that it is necessary to train large SOMs with a large number of genes and it is desirable to be able to annotate expansive datasets. The approach described in this paper utilises square SOMs with the training and computational time increasing quadratically as the area of the SOM is increased. This makes the use of the SOM impractical on serial machines when the SOM size is large e.g. training of a relatively small SOM (50x50) with 25000 genes for 3000 epochs using a single processor was measured as taking over 16 hours. The parallelisation method outlined here is ideally suited to the training of large SOMs. As the size of the SOM increases the benefits of parallel processing of SOM segments make the MPI communications overhead less significant. Cache and local memory efficiencies [11] are other contributors to increased performance for large SOMs. Segmenting the map into smaller processes also allows each SOM segment to reside in the cache of each processor. The cache has lower access latency than the RAM attached to each processor or the shared memory so the relative speedup for larger SOMs is increased. The reduced significance of the MPI overhead and cache efficiencies leads to excellent speed-ups for large SOMs. The speedups are illustrated in Fig. 3.

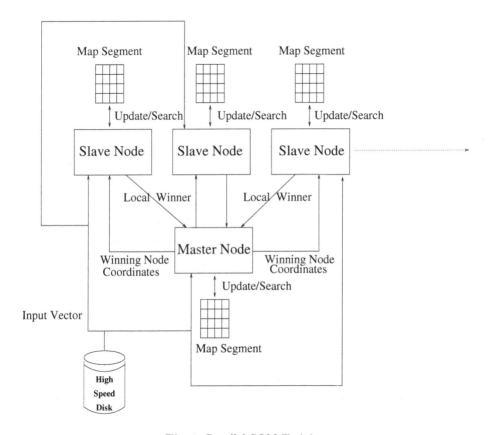

Fig. 1. Parallel SOM Training

5   Annotation

The first step was to create datasets that could be used for verification and testing of our SOM based annotation methods. A large collection of unannotated genomes were downloaded as well as the corresponding annotated versions from GenBank and a series of artificial metagenomes created with user defined mean read lengths and sequencing coverage. Subsets of known genes from the annotated genomes are used as training sets.

```
>gi|16445345|gb|AE007869.1| Agrobacterium tumefaciens str. C58 circular chromosome, complete sequence
::: read_num ::: 0 ::: read_length ::: 713 ::: gene_tag ::: genome2 ::: start_point ::: 1147449
TCGTCGCCTTTTATCTGCTGCTCGACTGGGACCGGATGATCGACAAGGTCGA
CAGCTGGATCCCGCGTGACTACGTGCATACCGTCCGCCAGATTGCTCGCGACATGGACAAGACGATTGCC
GGCTTCGTGCGCGGGGCAGGGGTCGCTCTGCCTCATTCTCGGGGTTTATTATGCCGTCGGCCTGTCGTTGG
TGGGCCTGAATTTCGGTCTGCTGATCGGCCTTTTCTCCGGCCTCATCAGCTTCATTCCCTATATCGGCTC
CATGGTCGGCCTCATCCTCGGCGGTCGGCGTCGGCCATCGTGCAGTTCGTGCCGGACTATATCTATGTCTTC
CTGACGGCTCGTCGTGTTCTTTTCCGGCCAGTTCATCGAAGGCAACATCCTGCAACCGAAGCTTGTCGGCA
AAAGCGTCGGCCTGGCATCCGGTCTGGCTGATGTTCGCGGCTCTTCGCTTTCGGTGCGCTTTTCGGTTTCGT
CGGCCTTCTGGTCGGCGGTTCCCCGGCTGCCGCTGCGGGTCGGCGTGCTTGTCGGCGTTTGCATTGTCGGCGGTAT
CTTGAAAGCCGATCTTTACCATGGGCACGCCGCCAATCTTCCCTGGGAAGCCAACCCTGCCCTGGAAGAAC
GGGAACCGTCCGCTGGTAAGATTGACTCTGCAGCCTGACGATGACTGACCAAATCAAAACCGACAACGCC
CGGTCGAAGGCCGAGCAGCTG
>gi|50082967|gb|AE017334.2| Bacillus anthracis str. 'Ames Ancestor', complete genome
::: read_num ::: 1 ::: read_length ::: 692 ::: gene_tag ::: genome3 ::: start_point ::: 639981
ACGATTTAATG
AAATCATTCGGTACTACGTTATTATTGTCCCTGTTATCTCGTGGCTTCAGTTCCGGTCGGCTCGGTCCGATTGGGT
TTATTGGTATTGTTACCCCGCATTTTGCCAGATTTCTTGTTGGGGTCGATCACAGATGGAGAGTACCGTA
TAGTGGTTTGTTAGGTGCAATACTATTAATCTTAGCTGATATAGCAGCAAGATATGTCATTATGCCCACAA
GAAGTACCAGTAGGGGTTATGACAGCATTTATTTGGGGCCCCATTCTTTATTTATTATTGCACGTAAGAGAG
GGCTTAGCAAATGAAGAAGTATATTCCGTTCCGTATAGGAAAAGGCGAGCTCTCGTTTTTAATGTATAAA
CGAGCTTGTCTCGTCTTTATTCAGTTACTAGTTGTTTTAATTGGATTATTTTTTGCCAGTGCAGGTATGG
GAGACATGAAAATTGCTCCTTATGATGTATGGCAAGCGATTACTGGAAATGGCGATGCGATGTCAAATAT
GGTTGTAAATAAGTTTCGTATGCCTCGTATTTTAATTGCTATCCTTGTAGGAATCGCACTTGCTGTAGCA
GGATGTATTTTACAAGGGCTCGTTCGAAATCCACTTGCTTCTCCGGATATCATCGGGATTACAGGTGGTG
CAAGTGTTGCAGTTGTATTATTTTTAGCTATATTTAGTGA
>gi|52426793|gb|CP000010.1| Burkholderia mallei ATCC 23344 chromosome 1, complete sequence
::: read_num ::: 2 ::: read_length ::: 800 ::: gene_tag ::: genome4 ::: start_point ::: 719409
TCGGCTATGGCTTTGACGTCACCGAGCACTGGA
CGGTATCGGCGAGCTACTCGAGCGCGCTTTCGCGCGCCGAGCTTCAACGATCTGTACTATCCGAACTCGGG
CAACCTGTCGATCCGCCCCGAGCGCAGCCATTCGGTCGAGCGGGCCGTGCAGTATGCGTCGGACGACGTG
GGCGTCGTGCGCGGTGACCGCGTTCCAGACCCGCTACACGAACCTGATCGACTATCGTCCGGCGCGGAGCG
GCCCTTACTATGTCGGCGCAGAACGTCGGCCGCGCGGAAGGTGCAGGGCGTCGAGGGATCGTGGCAGGGGCA
CGTCGGCGGGCACCGACGTGCCGCATTGCCGCGACGCTGCAGAACCCGGTCAACGAAACCGCCGGCCGCGAC
CTGAACCGGCGCGCGCGGCGGCTTCGCGTCGATCTCGGCGAACCGCTCGTTCGGCGGCTGGCGCGTCGGCG
```

Fig. 2. Snapshot of the metagenome

SOMs are trained using these multiple gene models and these SOMs are then used to predict genes within the metagenomic sequence reads. The parallelisation allows for the use of large datasets and large SOMs and therefore the encapsulation of many different gene coding models on the map. This is essential to identify the many various coding models that may exist within a metagenomic library. Parallel implementation also enables faster annotation of extremely large datasets, which is another requirement of metagenomic analysis. The success of the parallelised algorithm's predictions is examined by comparing the location of a prediction within its originating genome to the locations of all the genes in the annotated version and calculating the percentage of each prediction which is actually coding.

The ability of the SOMs to detect genes is shown experimentally. To show the ability of the method to detect genes for which exact coding models are not available a 5 member metagenome, at 5× coverage was created and annotated using a SOM trained with 25000 genes derived from 45 genomes not present in the metagenome. The size of the SOM was set at the square root of the training set size divided by 10 as proposed in [6] i.e. 50x50. In the experiment the learning rate α was initially set to .9 and reduced to .1 over the course of the the the ordering stage which consists of 1000 epochs (one third of the total number of epochs). In the convergence phase (2000 epochs) the learning rate was reduced from .1 to .034. The neighbourhood size was set to the same size as the SOM lattice and reduced in size by one lattice node from all directions when

$$epoch\% \frac{ordering}{size} = 0,\qquad (9)$$

during the ordering stage, where *epoch* is the latest number of completed epochs, *ordering* is the number of epochs in the ordering stage (total epochs/3) and *size* is the length of one side of the SOM. The neighbourhood size was set to a minimum of 0 in the convergence stage i.e. only the winning node is altered during each epoch. All genes in the training set were randomly chosen with no discrimination or bias with respect to protein families. In this random selection, no attempt is made to ensure that particular genes are not selected more that once and no attempt is made to account for the appearance of gene alleles in the training set, making the process a true representation of the shotgun sequencing process.

The benefits of the parallel training are evident when training SOMs even of this relatively small size. Training on 8 processors can be completed in just over 4 hours compared to 16 hours on a single processor and annotation can be completed in under 100 hours as compared to over 700 hours. The results of the annotation are seen in Table 1. E.g. 48% of reads that were predicted as being coding with a probability of .7 or above are 100% coding i.e. they lie completely within a gene.

6   Results

Parallelised training and annotation phases of the SOM formalism have been created and the necessary tools for creating metagenomes and multiple species training sets have been completed. The speedups for the parallelised training phase are shown in Fig 3. These speedups were measured on the Bull NovaScale NS6320 shared memory system belonging to the Irish Center for High End Computing (ICHEC). This machine comprises 32 Itanium 2 processors (running at 1.5GHz) and contains 256 GBytes of RAM. Linear speedups were seen in the

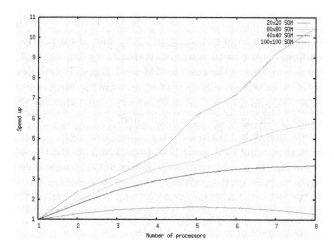

Fig. 3. Parallel Training Speedups - Training of large SOMs is required and a marked increase in performance for large SOMs is seen

annotation phase. It is these performance increases that enable analysis of the very large metagenomic datasets. Table 1 shows the results from an experimental annotation where the validity of predictions with a probability of .7 or above were examined. The annotation was carried out on a cluster consisting of 32 dual core Intel Xeon processors running at 2.4GHz.

Table 1. Results from a sample 5 member metagenome at 5× coverage using training genes not contained in the metagenome. Only predictions with a probability of .7 or above are examined. Here, the specificity is defined to be the percentage of each prediction that is confirmed as coding. An average sensitivity figure of 78% was recorded (meaning that 22% of the coding regions present in the metagenome were not found in any of the predictions). The results are expected to improve further with the use of larger numbers of training genes on large SOMs.

% of predictions	Specificity
48	100%
4	91% - 99%
10.4	81% - 90%
5.42	71% - 80%
4.52	61% - 70%
4.04	51% - 60%
3.71	41% - 50%
3.71	31% - 40%
5.19	21% - 30%
4.28	11% - 20%

7   Conclusions

The SOM neural network formalism has been introduced and has been shown to be applicable in genome annotation [6]. It has been shown that the SOM can be a useful algorithm in annotating metagenomic libraries without the need for creating scaffolds from metagenomic reads. The advantage of the multiple genes models used in the SOM over HMM and BLAST techniques has been identified. The computational challenges of annotating environmental samples have been highlighted and a proposed method to parallelising the SOM approach to overcome these challenges has been explained.

Significant speedups in SOM training times and metagenome annotation have been demonstrated proving that a HPC approach makes the use of the SOM in metagenomic analysis viable. Experiments using SOMs trained with large numbers of genes to annotate artificially generate metagenomes, created to simulate the process of WGS have been detailed and the results listed. With sufficient computing resources the approach could be used, with training sets consisting of hundreds of thousands of genes, so as to be representative of all the genes in GenBank aiding annotation of real environmental samples such as the Sargasso sea dataset and acid mine biofilm library through either direct gene prediction or by expediting sequence assembly through extraction of coding sequences.

References

1. Hugenholtz P. 2002. Exploring prokaryotic diversity in the genomic era. *Genome Biol 3:REVIEW0003*

2. Rappe, M., Giovanni, S. 2003. The uncultured microbial majority. *Annu Rev microbial 57*, pp. 369–394

3. Venter, J, C., Remington, K, Heidleberg, J. Environmental whole genome shotgun sequencing: The Sargasso Sea *Science 2004*

4. Chen, K. and Pachter, A. 2005. Bioinformatics for Whole-Genome Shotgun Sequencing of Microbial Communities. *PLoS Computational Biology 1:e24*

5. Riesenfeld, S, Christian., Schloss, D, Patrick. and Handelsman, Jo. 2004. Metagenomics: Genomic Analysis of Microbial Communities. *Annual Review of Genetics 38* pp. 525–552

6. Mahony, S., McInerny, J. O., Smith, T. J., Golden, A. 2004. Gene prediction using the Self Organising Map: automatic generation of multiple gene models *BMC Bioinformatics 5:23*

7. Abe, et al. 2003. Informatics for unveiling hidden genome signatures. *Genome Res. 13:* pp. 693–702

8. Kohonen, T. Self-Organizing Maps. *Berlin, Springer-Verlag 1995.*

9. Tomsich, P., Rauber, A., Merkl, D. 2000. Optimizing the parSOM neural network implementation for data mining with distributed memory systems and cluster computing *Proceedings. 11th International Workshop on Database and Expert Systems Applications* pp. 661–665

10. Snir, M., Otto, S., Huss-Lederman, S., Walker, D., and Dongarra, J. 1996 MPI: The Complete Reference *MIT Press, Cambridge, MA.*

11. Rauber, A., Tomsich, P., Merkl, D. 2000. parSOM: a parallel implementation of the self-organizing map exploiting cache effects: making the SOM fit for interactive high-performance data analysis *Proceedings of the IEEE-INNS-ENNS International Joint Conference on Neural Networks, Volume 6, pp 177–182*

A Distributed System for Genetic Linkage Analysis[*]

Mark Silberstein, Dan Geiger, and Assaf Schuster

Technion – Israel Institute of Technology
{marks | dang | assaf}@cs.technion.ac.il

Abstract. Linkage analysis is a tool used by geneticists for mapping disease-susceptibility genes in the study of Mendelian and complex diseases. However analyses of large inbred pedigrees with extensive missing data are often beyond the capabilities of a single computer. We present a distributed system called SUPERLINK-ONLINE for computing multipoint LOD scores of large inbred pedigrees. It achieves high performance via efficient parallelization of the algorithms in SUPERLINK, a state-of-the-art serial program for these tasks, and through utilization of thousands of resources residing in multiple opportunistic grid environments. Notably, the system is available online, which allows computationally intensive analyses to be performed with no need for either installation of software, or maintenance of a complicated distributed environment. The main algorithmic challenges have been to efficiently split large tasks for distributed execution in a highly dynamic non-dedicated running environment, as well as to utilize resources in all the available grid environments. Meeting these challenges has provided nearly interactive response time for shorter tasks while simultaneously serving massively parallel ones. The system, which is being used extensively by medical centers worldwide, achieves speedups of up to three orders of magnitude and allows analyses that were previously infeasible.

1 Introduction

Linkage analysis aims at facilitating the understanding of mechanisms of genetic diseases via identification of the areas on the chromosome where disease-susceptibility genes are likely to reside. Computation of a logarithm of odds (LOD) is a valuable tool widely used for the analysis of disease-susceptibility genes in the study of Mendelian and complex diseases. The computation of the LOD score for large inbred pedigrees with extensive missing data is often beyond the computation capabilities of a single computer. Our goal is to facilitate these more demanding linkage computations via parallel execution on multiple computers.

We present a distributed system for exact LOD score computations, called SUPERLINK-ONLINE [1], capable of analyzing inbred families of several hundreds individuals with extended missing data using single and multiple loci disease models. Our approach, made possible by recent advances in distributed computing (e.g., [2]), eliminates the need for expensive hardware by distributing the computations over thousands of non-dedicated PCs and utilizing their idle cycles. Such opportunistic environments,

---

[*] This work is supported by the Israeli Ministry of Science.

W. Dubitzky et al. (Eds.): GCCB 2006, LNBI 4360, pp. 110–123, 2007.

referred further as *grids*, are characterized by the presence of many computers with different capabilities and operating systems, frequent failures, and extreme fluctuations in the number of computers available. Our algorithm allows a single linkage analysis task to be recursively split into multiple independent subtasks of required size; each subtask can be further split into smaller subtasks for performance reasons. The flexibility to adjust the number of subtasks to the amount of available resources is particularly important in grids. The subtasks are then executed on grid resources in parallel, and their results are combined and outputted as if executed on a single computer.

SUPERLINK-ONLINE delivers the newest information technology to geneticists via a simple Internet interface, which completely hides the complexity of the underlying distributed system. The system allows for concurrent submission and execution of linkage tasks by multiple users, generating a *stream* of parallel and serial tasks with vastly different (though unknown in advance) computational requirements. These requirements range from a few seconds on a single CPU for some tasks to a few days on thousands for others. The stream is dynamically scheduled on a set of independently managed grids with vastly different amounts of resources. In addition, the time spent in a grid on activities other than actual task execution differs for different grids, and the difference can reach several orders of magnitude. We refer to such activities – such as resource allocation or file transfer – as *execution overhead*. The challenge is to provide nearly interactive response time for shorter tasks (up to a few seconds), while simultaneously serving highly parallel heavy tasks in such a multi-grid environment.

Our system implements a novel scheduling algorithm which combines the well-known Multilevel Feedback Queue (MQ) [3] approach with the new concept of *grid execution hierarchy*. All available grids in the hierarchy are sorted according to their size and overhead: upper levels of the hierarchy include smaller grids with faster response time, whereas lower levels consist of one or more large-scale grids with higher execution overhead. The task stream is scheduled on the hierarchy, so that each task is executed at the level that matches the task's computational requirements, or its *complexity*. The intuition behind the scheduling algorithm is as follows. As the task complexity increases, so do the computational requirements and the tolerable overhead. Consequently, more complex tasks should be placed at a lower level of the hierarchy, and parallelized according to the available resources at that level.

The proper execution level for a given task is easy to determine if the task complexity is known or simple to compute. Unfortunately, this is not the case for the genetic linkage analysis tasks submitted to the system. In fact, geneticists usually do not know whether the results can be obtained in a few seconds, or whether the task will require several years of CPU time. Provably, for the linkage analysis tasks estimating the exact complexity of a given task is NP-hard in itself [4].

Our algorithm allows the proper execution level to be quickly determined via fast estimation of an upper bound on the task complexity. We apply a heuristic algorithm that yields such an upper bound [5] for a fraction of the running time limit assigned to a given hierarchy level. If the complexity is within the complexity range of that level (which is configured before the system is deployed), the task is executed; otherwise it is moved directly to a lower level. The precision of the complexity estimation algorithm improves the longer it executes. Consequently, task complexity is reassessed at each level prior to the actual task execution. The lower the level, the greater the amount of resources that are allocated for estimating more precisely its complexity, resulting in a better matched hierarchy level.

SUPERLINK-ONLINE is a production system which assists geneticists worldwide, utilizing about 3000 CPUs located in several grids at the Technion in Haifa, and at the University of Wisconsin in Madison. During the last year the system served over 7000 tasks, consuming about 110 CPU years over all available grids.

The paper is organized as follows. In Section 2 we show the algorithm for scheduling a stream of linkage analysis tasks on a set of grids. In Section 3 we describe the parallel algorithm for genetic linkage analysis. The current deployment of the SUPERLINK-ONLINE system is reported in Section 4, followed by performance evaluation, related work and future research directions.

2   The Grid Execution Hierarchy Scheduling Algorithm

The algorithm has two complementary components: organization of multiple grids into a grid execution hierarchy and a set of procedures for scheduling tasks on this hierarchy.

2.1   Grid Execution Hierarchy

The purpose of the execution hierarchy is to classify available grids according to their performance characteristics, such as execution overhead and amount of resources, so that each level of the hierarchy provides the best performance for tasks of a specific complexity range.

Upper levels of the hierarchy include smaller grids with faster response time, whereas lower levels consist of one or more large-scale grids with higher execution overhead. The number of levels in the hierarchy depends on the expected distribution of task complexities in the incoming task stream, as explained in Section 4.

Each level of the execution hierarchy is associated with a set of one or more queues. Each queue is connected to one or more grids at the corresponding level of the hierarchy, allowing submission of jobs into these grids. A task arriving at a given hierarchy level is enqueued into one of the queues. It can be either executed on the grids connected to that queue (after being split into jobs for parallel execution), or migrated to another queue at the same level by employing simple load balancing techniques. If a task does not match the current level of the execution hierarchy, as determined by the scheduling procedure presented in the next subsection, it is migrated to the queue at a lower level of the hierarchy.

2.2   Scheduling Tasks in Grid Hierarchy

The goal of the scheduling algorithm is to quickly find the proper execution hierarchy level for a task of a given complexity . Ideally, if we knew the complexity of each task and the performance of each grid in the system, we could compute the execution time of a task on each grid, placing that task on the one that provides the shortest execution time. In practice, however, neither the task complexity nor the grid performance can be determined precisely. Thus, the algorithm attempts to schedule a task using approximate estimates of these parameters, dynamically adjusting the scheduling decisions if these estimates turn out to be incorrect.

We describe the algorithm in steps, starting with the simple version, which is then enhanced. Due to space limitations we omit some details available elsewhere [6].

Simple MQ with grid execution hierarchy. Each queue in the system is assigned a maximum time that a task may execute in the queue (execution time)T_e[1]. The queue configured to serve the shortest tasks is connected to the highest level of the execution hierarchy, the queue for somewhat longer tasks is connected to the next level, and so on.

A task is first submitted to the top level queue. If the queue limit T_e is violated, a task is preempted, check-pointed and restarted in the next queue (the one submitting tasks to the next hierarchy level).

Such an algorithm ensures that any submitted task will eventually reach the hierarchy level that provides enough resources for longer tasks and fast response time for shorter tasks. In fact, this is the original MQ algorithm applied to a grid hierarchy.

However, the algorithm fails to provide fast response time to short tasks if a long task is submitted to the system prior to a short one. Recall that the original MQ is used in time-shared systems, and tasks within a queue are scheduled using preemptive round-robin, thus allowing fair sharing of the CPU time [3]. In our case, however, tasks within a queue are served in FCFS manner (though later tasks are allowed to execute if a task at the head of the queue does not occupy all available resources). Consequently, a long task executed in a queue for short tasks may make others wait until its own time limit is exhausted.

Quick evaluation of the expected running times of a task in a given queue can prevent the task from being executed at the wrong hierarchy level. Tasks expected to consume too much time in a given queue are migrated to the lower level queue without being executed. This is described in the next subsection.

Avoiding hierarchy level mismatch. Each queue is associated with a maximum allowed task complexity C_e, derived from the maximum allowed execution time T_e by optimistically assuming linear speedup, availability of all resources at all times, and resource performance equal to the average in the grid. The optimistic approach seems reasonable here, because executing a longer task in an upper level queue is preferred over moving a shorter task to a lower level queue, which could result in unacceptable overhead. The following naive relationship between a complexity limit C_e and a time limit T_e reflects these assumptions:

$$C_e = T_e \cdot (N \cdot P \cdot \beta), \qquad (1)$$

where N is the maximum number of resources that can be allocated for a given task, P is the average resource performance, and β is the efficiency coefficient of the

---

[1] For simplicity we do not add the queue index to the notations of queue parameters, although they can be set differently for each queue.

application on a single CPU, defined as a portion of CPU-bound operations in the overall execution.

For each task arriving to the queue, task's complexity is first estimated[2]. Allocating a small portion $\alpha < 1$ of T_e for complexity estimation often allows quick detection of a task that does not fit in the queue. The task is migrated to the lower level queue if its complexity estimate is higher than C_e.

However, the upper bound on the task's complexity might be much larger than the actual value. Consequently, if the task is migrated directly to the level in the hierarchy that serves the complexity range in which this estimate belongs, it may be moved to too low a level, thus decreasing the application's performance. Therefore, the task is migrated to the next level, where the complexity estimation algorithm is given more resources and time to execute, resulting in a more precise estimate.

The queue complexity mismatch detection improves the performance of both shorter and longer tasks. On the one hand, the computational requirements of longer tasks are quickly discovered, and the tasks are moved to the grid hierarchy level with enough computational resources avoiding the overhead of the running task migration. On the other hand, longer tasks neither delay nor compete with the shorter ones, reducing the response time and providing nearly interactive user experience. In practice, however, a task may stay in the queue longer than initially predicted because of the too optimistic grid performance estimates. Thus, the complexity mismatch detection is combined with the enforcement of the queue time limit described in the previous subsection, by periodically monitoring the queue and migrating tasks which violate the queue time limit.

2.3   Handling Multiple Grids at the Same Level of the Execution Hierarchy

The problem of scheduling in such a configuration is equivalent to the well-studied problem of load sharing in multi-grids. It can be solved in many ways, including using the available meta-schedulers, such as [7], or flocking [2]. If no existing load sharing technologies can be deployed between the grids, we implement load sharing as follows.

Our implementation is based on a push migration mechanism (such as in [8]) between queues, where each queue is connected to a separate grid. Each queue periodically samples the availability of resources in all grids at its level of the execution hierarchy. This information, combined with the data on the total workload complexity in each queue, allows the expected completion time of tasks to be estimated. If the current queue is considered suboptimal, the task is migrated. Conflicts are resolved by reassessing the migration decisions at the moment the task is moved to the target queue. Several common heuristics are implemented to reduce sporadic migrations that may occur as a result of frequent fluctuations in grid resource availability [9]. Such heuristics include, for example, averaging of momentary resource availability data with the historical data, preventing migration of tasks with a small number of pending execution requests, and others.

---

[2] The complexity estimation can be quite computationally demanding for larger tasks, and in which case it is executed using grid resources.

3 Parallel Algorithm for the Computation of LOD Score

LOD score computation can be represented as the problem of computing an expression of the form

$$\sum_{x_1} \sum_{x_2} \cdots \sum_{x_n} \prod_{i=1}^{m} \Phi_i(\mathbf{X_i}), \qquad (2)$$

where $\mathbf{X} = \{x_1, x_2, \ldots, x_n | x_i \in \mathbb{N}\}$ is a set of non-negative discrete variables, $\Phi_i(\mathbf{X_i})$ is a function $\mathbb{N}^k \rightarrow \mathbb{R}$ from the subset $\mathbf{X_i} \subset \mathbf{X}$ of these variables of size k to the reals, and m is the total number of functions to be multiplied. Functions are specified by a user as an input. For more details we refer the reader to [10].

3.1 Serial Algorithm

The problem of computing Eq. 2 is known to be NP-complete [11]. One possible algorithm for computing this expression is called variable elimination [12].

The complexity of the algorithm is fully determined by the order in which variables are eliminated. For example, two different orders may result in running time ranging from few seconds to several hours. Finding an optimal elimination order is NP-complete [13]. A close-to-optimal order can be found using the stochastic greedy algorithm proposed in [5]. The algorithm can be stopped at any point, and it produces better results the longer it executes, converging faster for smaller problems. The algorithm yields a possible elimination order and an upper bound on the problem complexity, i.e., the number of operations required to carry out the computations if that order is used. It is this feature of the algorithm that is used during the scheduling phase to quickly estimate the complexity of a given problem prior to execution.

The above algorithm is implemented in SUPERLINK, which is the state-of-the-art program for performing LOD score computations of large pedigrees.

3.2 Parallel Algorithm

The choice of a parallelization strategy is guided by two main requirements. First, sub-tasks are not allowed to communicate or synchronize their state during the execution. This factor is crucial for performance of parallel applications in a grid environment, where communication between computers is usually slow, or even impossible due to security constraints. Second, parallel applications must tolerate frequent failures of computers during the execution by minimizing the impact of failures on the overall performance.

The algorithm for finding an elimination order consists of a large number of independent iterations. This structure of the algorithm allows to parallelize it by distributing the iterations over different CPUs using the master-worker paradigm, where master process maintains a queue of independent jobs which are subsequently pulled and executed by worker processes running in parallel on multiple CPUs. At the end of the execution the order with the lowest complexity is chosen.

Parallelization of the variable elimination algorithm also fits the master-worker paradigm and is performed as follows. We represent the first summation over x_1 in

Eq.2 as a sum of the results of the remaining computations, performed for every value of x_1. This effectively splits the problem into a set of independent subproblems having exactly the same form as the original one, but with the complexity reduced by a factor approximately equal to the number of values of x_1. We use this principle recursively to create subproblems of a desired complexity, increasing the number of variables used for parallelization. Each subproblem is then executed independently, with the final result computed as the sum of all partial results.

The choice of the number of subtasks and their respective granularity is crucial for efficiency of the parallelization. The inherent overheads of distributed environments, such as scheduling and network delays, often become a dominating factor inhibiting meaningful performance gains, suggesting that long-running subtasks should be preferred. On the other hand, performance degradation as a result of computer failures will be lower for short subtasks, suggesting to reduce the amount of computations per subtask. Furthermore, decreasing the amount of computations per subtask increases the number of subtasks generated for computing a given problem, improving load balancing and utilization of available computers.

Our algorithm controls the subtask granularity by specifying maximum allowable complexity threshold C. Specifying lower values of C increases the number of subtasks, decreasing the subtask complexity and consequently its running time. The value of C for a given problem is determined as follows. We initially set C so that a subtask's running time does not exceed the average time it can execute without interruption on a computer in the Grid environment being used. If such value of C yields that the number of subtasks is below the number of available computers, then C is iteratively reduced to allow division into more subtasks. The lower bound on C is set so that overheads due to scheduling and network delays constitute less than 1% of the subtask's running time.

3.3   Implementation of the Parallel Algorithm for Execution in Grid Hierarchy

Parallel tasks are executed in a distributed environment via Condor [2], which is a general purpose distributed batch system, capable of utilizing idle cycles of thousands of CPUs[3]. Condor hides most of the complexities of job invocation in an opportunistic environment. In particular, it handles job failures that occur because of changes in the system state. Such changes include resource failures, or a situation in which control of a resource needs to revert to its owner. Condor also allows resources to be selected according to the user requirements via a matching mechanism.

There are three stages in running master-worker applications in Condor: the parallelization of a task into a set of independent jobs, their parallel execution via Condor, and the generation of final results upon their completion. In our implementation, this flow is managed by the Condor flow execution engine, called DAGman, which invokes jobs according to the execution dependencies between them, specified as a directed acyclic graph (DAG). The complete genetic linkage analysis task comprises two master-worker applications, namely, parallel ordering estimation and parallel variable elimination. To integrate these two applications into a single execution flow, we use an outer DAG composed of two internal DAGs, one for each parallel application.

---

[3] Use of Condor is nor assumed, neither required by the algorithm. Moreover, the system is being expanded to use other grid batch systems without making any change to the algorithm.

DAGman is capable of saving a snapshot of the flow state, and then restarting execution from this snapshot at a later time. We use this feature for migration of a task to another queue as follows: the snapshot functionality is triggered, all currently executing subtasks are preempted, the intermediate results are packed, and the task is transferred to another queue where it is restarted.

4   Deployment of SUPERLINK-ONLINE

The current deployment of the SUPERLINK-ONLINE portal is presented in Figure 1.

We configure three levels of the execution hierarchy, for the following reasons. About 60% of the tasks take a few minutes or less, and about 28% take less than three hours, as reflected by the histogram in Figure 4, generated from the traces of about 2000 tasks executed by the system. This suggests that two separate levels, Level 1 and Level 2, should be allocated for these dominating classes, leaving Level 3 for the remaining longer tasks. Yet, the current system is designed to be configured with any number of levels to accommodate more grids as they become available.

Each queue resides on a separate machine, connected to one or several grids. Utilization of multiple grids via the same submission machine is enabled by the Condor flocking mechanism, which automatically forwards job execution requests to the next grid in the list of available (*flocked*) grids, if these jobs remain idle after previous resource allocation attempts in the preceding grids.

Queue Q1 is connected to the dedicated dual CPU server and invokes tasks directly without parallelization. Queue Q2 resides on a submission machine connected to the flock of two Condor pools at the Technion. However, due to the small number of resources at Q2, we increased the throughput at Level 2 of the execution hierarchy by activating load sharing between Q2 and Q3, which is connected to the flock of three Condor pools at the University of Wisconsin in Madison. The tasks arrive to Q3 only from Q2. Queue Q4 is also connected to the same Condor pools in Madison as Q3, and may receive tasks from both Q2 and Q3.

In fact, Q3 exhibits rather high invocation latencies and does not fit Level 2 of the execution hierarchy well. Alternatively, Q3 could have been set as an additional level between Q2 and Q4, and the queue time limit of Q2 could have been adjusted to handle smaller tasks. However, because both queues can execute larger tasks efficiently, such partitioning would have resulted in unnecessary fragmentation of resources. Migration allows for a more flexible setup, which takes into account load, resource availability and overheads in both queues, and moves a whole task from Q2 to Q3 only if this yields better task performance. Typically, small tasks are not migrated, while larger tasks usually benefit from execution in a larger grid as they are allocated more execution resources. This configuration results in better performance for smaller tasks than does flocking between these two grids, as it ensures their execution on the low-overhead grid.

To ensure that larger tasks of Q4 do not delay smaller tasks of Q3, jobs of tasks in Q3 are assigned higher priority and may preempt jobs from Q4. Starvation is avoided via internal Condor dynamic priority mechanisms [2].

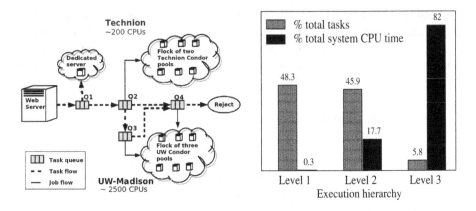

Fig. 1. SUPERLINK-ONLINE deployment

Fig. 2. Tasks handled versus the overall system CPU time (per level)

5   Results

We evaluate two different aspects of the performance of the SUPERLINK-ONLINE system. First, we show the speedups in the running time achieved by the parallel algorithm implementation over the serial state-of-the-art program for linkage analysis on a set of challenging inputs. Next, we demonstrate the performance of the scheduling algorithm by analyzing the traces of about 2,000 tasks served by SUPERLINK-ONLINE during second half of 2005.

5.1   Parallel Algorithm Performance

We measured the total run time of computing the LOD score for a few real-life challenging inputs. The results reflect the time a sole user would wait for a task to complete, from submission to the SUPERLINK-ONLINE system via a web interface until an e-mail notification about task completion is sent back. Results are summarized in Table 1.

We also compared the running time with that of the newest serial version of SU-PERLINK, invoked on Intel Pentium Xeon 64bit 3.0 Ghz, 2GB RAM. The entries of the running time exceeding two days were obtained by measuring the portion of the problem completed within two days, as made available by SUPERLINK, and extrapolating the running time assuming similar progress. The time saving of the online system versus a single computer ranged from a factor of 10 to 700. The results (in particular, rows 5,7,9,10) demonstrate the system capabilities to perform the analyses which are infeasible by SUPERLINK, considered a state-of-the-art program for linkage analysis on such inputs.

In a large multiuser Grid environment used by SUPERLINK-ONLINE, the number of computers employed in computations of a given task may fluctuate during the execution between only few to several hundreds. Table 1 presents the average and the maximum number of computers used during execution. We believe that the performance can be improved significantly if the system is deployed in a dedicated environment. Running times of SUPERLINK-ONLINE include network delays and resource failures (handled

Table 1. SUPERLINK-ONLINE vs. serial SUPERLINK V1.5

Input	Running time		#CPU used	
	SUPERLINK V1.5	SUPERLINK-ONLINE	Average	Maximum
1	5000sec	1050sec	10	10
2	5600sec	520sec	11	11
3	20hours	2hours	23	30
4	450min	47min	82	83
5	~300hours	7.1hours	38	91
6	297min	27min	82	100
7	~138days	6.2hours	349	450
8	~2092sec	1100sec	7	8
9	~231hours	3hours	139	500
10	~160days	8hours	310	360

automatically by the system). The column average is computed from the number of computers sampled every 5 minutes during the run.

Example of a previously infeasible analysis. We replicated the analysis performed for studying the Cold Induced Sweating Syndrome in a Norwegian family [14]. The pedigree consists of 93 individuals, two of which affected, and only four were typed. The original analysis was done using FASTLINK [15]. The maximum LOD score of 1.75 was reported using markers D19S895, D19S566 and D19S603, with the analysis limited to only three markers due to computational constraints[4]. According to the authors, using more markers for the analysis was particularly important in this study as "*in the absence of ancestral genotypes, the probability that a shared segment is inherited IBD from a common ancestor increases with the number of informative markers contained in the segment; that is, a shared segment containing only three markers has a significant probability of being simply identical by state, whereas a segment containing a large number of shared markers is much more likely to be IBD from a common ancestor. Thus, the four-point LOD score on the basis of only three markers from within an interval containing 13 shared markers, is an underestimate of the true LOD score*". Study of another pedigree as well as application of additional statistical methods were employed to confirm the significance of the findings.

Using SUPERLINK-ONLINE we computed six point LOD score with the markers D19S895, M4A, D19S566, D19S443 and D19S603, yielding the LOD=3.10 at marker D19S895, which would facilitate the linkage conclusion based on the study of this pedigree alone.

5.2   Performance of the Grid Execution Hierarchy Scheduling Algorithm

We analyzed the traces of 2300 tasks, submitted to the SUPERLINK-ONLINE system via Internet by users worldwide for the period between the 1st of June and the 31st of December 2005. During this time, the system utilized about 460,000 CPU hours (52.5

---

[4] LOD score above 3.3 is considered an indication of linkage.

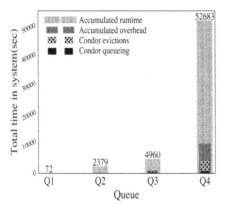
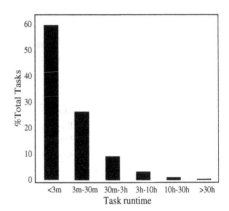

Fig. 3. Average accumulated time (from arrival to termination) of tasks in each queue

Fig. 4. Distribution of real task runtimes as registered by the system

CPU years) over all Condor pools connected to it (according to the Condor accounting statistics). This time reflects the time that would have been spent if all tasks were executed on single CPU. About 70% of the time was utilized by 1971 successfully completed tasks. Another 3% was wasted because of system failures and user-initiated task removals. The remaining 27% of the time was spent executing tasks which failed to complete within the queue time limit of Q4, and were forcefully terminated. However, this time should not be considered as lost since users were able to use partial results. Still, for clarity, we do not include these tasks in the analysis.

We compared the total CPU time required to compute tasks by each level of the execution hierarchy relative to the total system CPU consumption by all levels together. As expected, the system spent most of its time handling the tasks at Level 3, comprising 82% of the accumulated running time of the system (see Figure 2). The tasks at Level 2 consumed only 17.7% of the total system bandwidth, and only 0.3% of the time was consumed by the tasks at Level 1. If we consider the total number of tasks served by each level, the picture is reversed: the first two levels served significantly more tasks than the lower level. This result proves that the system was able to provide short response time to the tasks which were served at the upper levels of the hierarchy.

This conclusion is further supported by the graph in Figure 3, which depicts the average accumulated time of tasks in each queue, computed from the time a task is submitted to the system until it terminates. This time includes accumulated overheads, which are computed by excluding the time of actual computations from the total time. As previously, the graph shows only the tasks which completed successfully. Observe that very short tasks which require less than three minutes of CPU time and are served by Q1 stay in the system only 72 seconds on average regardless of the load in the other queues. This is an important property of the scheduling algorithm.

The graph also shows average accumulated overhead for tasks in each queue, which is the time a task spent in the system performing any activity other than the actual computations.

The form of the graph requires explanation. Assuming a uniform distribution of task runtimes and availability of an appropriate grid for each task, the task accumulated time is expected to increase linearly towards lower levels of the hierarchy. However in practice these assumptions do not hold. There are exponentially more shorter tasks requiring up to 3 minutes on single CPU (see Figure 4). This induces a high load on Q1, forcing short tasks to migrate to Q2 and thus reducing the average accumulated time of tasks in Q2. This time is further reduced by the load sharing between Q2 and Q3, which causes larger tasks to migrate from Q2 to Q3. Thus, shorter tasks are served by Q2, while longer ones are executed in Q3, resulting in the observed difference between the accumulated times in these queues. To explain the observed steep increase in the accumulated time in Q4, we examined the distribution of running times in this queue. We found that shorter tasks (while exceeding Q3's allowed task complexity limit) were delayed by longer tasks that preceded them. Indeed, over 70% of the overhead in that queue is due to the time the tasks were delayed because of other tasks executing in that queue. Availability of additional grids for the execution of higher complexity tasks would allow for the queueing and turnaround time to be reduced.

6   Related Work and Conclusions

In this work we presented the SUPERLINK-ONLINE system, which delivers the power of grid computing to geneticists worldwide, and allows them to perform previously infeasible analyses in their quest for disease-provoking genes. We described our parallel algorithm for genetic linkage analysis, suitable for execution in grids, and the method for organizing multiple grids and scheduling user-submitted linkage analysis tasks in such a multi-grid environment.

Future work will include on-the-fly adaptation of the hierarchy to the changing properties of the grids, designing a cost model to take into account locality of applications and execution platforms, improving the efficiency of the parallel algorithm, as well as connection to other grids, such as EGEE-2.

Related work. Parallelization of linkage analysis has been reported in [15–22]. However, the use of these programs by geneticists has been quite limited due to their dependency on the availability of high performance execution environments, such as a cluster of high performance dedicated machines or a supercomputer, and due to their operational complexity.

Execution of parallel tasks in grid environments has been thoroughly studied by grid researchers (e.g., [23–28]). In particular, [28] addressed the problem of resource management for short-lived tasks on desktop grids, demonstrating the suitability of grid platforms for execution of short-lived applications. Various scheduling approaches have been reported in many works(e.g. meta-schedulers [7, 29][30, 31], streams of tasks scheduling [32], steady-state scheduling [33] and others), however some of the are either not applicable in a real environment due to unrealistic assumptions. Others, such as [9] and [8], inspired the implementation of some system components.

References

1. Superlink-online: Superlink-online genetic linkage analysis system. http://bioinfo.cs.technion.ac.il/superlink-online (2006)
2. Thain, D., Livny, M.: Building reliable clients and servers. In Foster, I., Kesselman, C., eds.: The Grid: Blueprint for a New Computing Infrastructure. Morgan Kaufmann, San-Francisco (2003)
3. Kleinrock, L., Muntz, R.: Processor sharing queueing models of mixed scheduling disciplines for time shared systems. Journal of ACM **19** (1972) 464–482
4. Fishelson, M., Geiger, D.: Exact genetic linkage computations for general pedigrees. Bioinformatics **18**(Suppl. 1) (2002) S189–S198
5. Fishelson, M., Dovgolevsky, N., Geiger, D.: Maximum likelihood haplotyping for general pedigrees. Human Heredity **59** (2005) 41–60
6. Silberstein, M., Geiger, D., Schuster, A., Livny, M.: Scheduling of mixed workloads in multi-grids: The grid execution hierarchy. In: 15th IEEE International Symposium on High Performance Distributed Computing (HPDC-15 2006). (2006)
7. CSF: Community scheduler framework. http://www.globus.org/toolkit/docs/4.0/contributions/csf (2006)
8. England, D., Weissman, J.: Costs and benefits of load sharing in the computational grid. In Feitelson, D.G., Rudolph, L., eds.: 10th Workshop on Job Scheduling Strategies for Parallel Processing. (2004)
9. Vadhiyar, S., Dongarra, J.: Self adaptivity in grid computing. Concurrency and Computation: Practice and Experience **17**(2-4) (2005) 235–257
10. Friedman, N., Geiger, D., Lotner, N.: Likelihood computation with value abstraction. In: Proceedings of the 16th Conference on Uncertainty in Artificial Intelligence (UAI), Morgan Kaufmann (2000) 192–200
11. Cooper, G.: The computational complexity of probabilistic inference using bayesian belief networks. Artificial Intelligence **42** (1990) 393–405
12. Dechter, R.: Bucket elimination: A unifying framework for probabilistic inference. In Jordan, M., ed.: Learning in Graphical Models. Kluwer Academic Press. (1998) 75–104
13. Arnborg, S., Corneil, D.G., Proskurowski, A.: Complexity of finding embeddings in a k-tree. SIAM Journal of Algorithms and Discrete Methods **8** (1987) 277–284
14. Knappskog, P., Majewski, J., Livneh, A., Nilsen, P., Bringsli, J., Ott, J., Boman, H.: Cold-Induced Sweating Syndrome is caused by mutations in the CRLF1 Gene. American Journal of Human Genetics **72(2)** (2003) 375–383
15. Miller, P., Nadkarni, P., Gelernter, G., Carriero, N., Pakstis, A., Kidd, K.: Parallelizing genetic linkage analysis: a case study for applying parallel computation in molecular biology. Computing and Biomedical Research **24**(3) (1991) 234–248
16. Dwarkadas, S., Schäffer, A., Cottingham, R., Cox, A., Keleher, P., Zwaenepoel, W.: Parallelization of general linkage analysis problems. Human Heredity **44** (1994) 127–141
17. Matise, T., Schroeder, M., Chiarulli, D., Weeks, D.: Parallel computation of genetic likelihoods using CRI-MAP, PVM, and a network of distributed workstations. Human Heredity **45** (1995) 103–116
18. Gupta, S., Schäffer, A., Cox, A., Dwarkadas, S., Zwaenepoel, W.: Integrating parallelization strategies for linkage analysis. Computing and Biomedical Research **28** (1995) 116–139
19. Rai, A., Lopez-Benitez, N., Hargis, J., Poduslo, S.: On the parallelization of Linkmap from the LINKAGE/FASTLINK package. Computing and Biomedical Research **33**(5) (2000) 350–364
20. Kothari, K., Lopez-Benitez, N., Poduslo, S.: High-performance implementation and analysis of the linkmap program. Computing and Biomedical Research **34**(6) (2001) 406–414

21. Conant, G., Plimpton, S., Old, W., Wagner, A., Fain, P., Pacheco, T., Heffelfinger, G.: Parallel Genehunter: implementation of a linkage analysis package for distributed-memory architectures. Journal of Parallel and Distributed Computing **63**(7-8) (2003) 674–682

22. Dietter, J., Spiegel, A., an Mey, D., Pflug, H.J., al Kateb, H., Hoffmann, K., Wienker, T., Strauch, K.: Efficient two-trait-locus linkage analysis through program optimization and parallelization: application to hypercholesterolemia. European Journal of Human Genetics **12** (2005) 542–50

23. Berman, F., Wolski, R.: Scheduling from the perspective of the application. In: 12th IEEE International Symposium on High Performance Distributed Computing (HPDC'03), Washington, DC, USA, IEEE Computer Society (1996) 100–111

24. Yang, Y., Casanova, H.: Rumr: Robust scheduling for divisible workloads. In: 12th IEEE International Symposium on High Performance Distributed Computing (HPDC'03), Washington, DC, USA, IEEE Computer Society (2003) 114

25. Berman, F., Wolski, R., Casanova, H., Cirne, W., Dail, H., Faerman, M., Figueira, S., Hayes, J., Obertelli, G., Schopf, J., Shao, G., Smallen, S., Spring, S., Su, A., Zagorodnov, D.: Adaptive computing on the grid using AppLeS. IEEE Transactions on Parallel and Distributed Systems **14(4)** (2003) 369–382

26. Heymann, E., Senar, M.A., Luque, E., Livny, M.: Adaptive scheduling for master-worker applications on the computational grid. In: GRID 2000. (2000) 214–227

27. Beaumont, O., Legrand, A., Robert, Y.: Scheduling divisible workloads on heterogeneous platforms. Parallel Computing **29**(9) (2003) 1121–1152

28. Kondo, D., Chien, A.A., Casanova, H.: Resource management for rapid application turnaround on enterprise desktop grids. In: ACM/IEEE Conference on Supercomputing (SC'04), Washington, DC, USA, IEEE Computer Society (2004) 17

29. MOAB Grid Suite: Moab grid suite. http://www.clusterresources.com/pages/ products/moab-grid-suite.php (2006)

30. Dail, H., Sievert, O., Berman, F., Casanova, H., YarKhan, A., Vadhiyar, S., Dongarra, J., Liu, C., Yang, L., Angulo, D., Foster, I.: Scheduling in the grid application development software project. Grid Resource Management: State-of-the-art and Future Trends (2004) 73–98

31. Vadhiyar, S., Dongarra, J.: A metascheduler for the grid. In: 11th IEEE International Symposium on High Performance Distributed Computing (HPDC'02), Washington, DC, USA, IEEE Computer Society (2002)

32. Sabin, G., Kettimuthu, R., Rajan, A., Sadayappan, P.: Scheduling of parallel jobs in a heterogeneous multi-site environment. In Feitelson, D.G., Rudolph, L., Schwiegelshohn, U., eds.: Job Scheduling Strategies for Parallel Processing. Springer Verlag (2003) 87–104

33. Marchal, L., Yang, Y., Casanova, H., Robert, Y.: Steady-state scheduling of multiple divisible load applications on wide-area distributed computing platforms. International Journal of High Performance Computing Applications (2006, to appear)

Enabling Data Sharing and Collaboration in Complex Systems Applications

Michael A. Johnston and Jordi Villà-Freixa

Computational Biochemistry and Biophysics Laboratory, Research Unit on
Biomedical Informatics (GRIB) Institut Municipal d'Investigació Mèdica and
Universitat Pompeu Fabra, C/Doctor Aiguader, 88 08003 Barcelona, Catalunya,
Spain

Abstract. We describe a model for the data storage, retrieval and manipulation requirements of complex system applications based on pervasive, integrated, application specific databases shared over a peer to peer
network. Such a model can significannotly increase productivity through
transparent data sharing and querying as well as aid collaborations. We
show a proof of concept of this approach as implemented in the Adun
molecular simulation application together with a discussion of its limitations and possible extensions.

1 Introduction

As stated in the NIH Data Sharing Policy and Implementation Guidance [1],
sharing data (sic.), "encourages diversity of analysis and opinion, promotes new
research, makes possible the testing of new or alternative hypotheses and methods of analysis, enables the exploration of topics not envisioned by the initial
investigators, and permits the creation of new datasets when data from multiple
sources are combined".

Data sharing is a particularly important issue for unique data that cannot
be readily replicated. Such data is prevalent in the study of complex systems
through computer simulation where data sets usually require massive computing
resources and long times to generate. Thus, although it is possible to run the
same simulation more than once this represents a huge squandering of resources.
It is also a valuable tool for collaborations where easy access to joint data,
including new results and analysis, enhances productivity.

1.1 Data Sharing Methods

There are four main data sharing technologies: data grids, peer to peer (P2P)
networks, distributed databases and content delivery networks (CDN). In the
context of sharing complex systems data only the first three need be considered (CDN's require special infrastructure and are usually propriety in nature)
and a brief introduction and comparison, along with references to examples of
implementations, is given here. A more detailed comparison of these methods,
focusing on data grids, can be found in [2].

W. Dubitzky et al. (Eds.): GCCB 2006, LNBI 4360, pp. 124–140, 2007.

Data Grids. Data grids focus on sharing distributed resources and supporting computations over the same infrastructure. They are created by institutions forming virtual organisations (VO) and pooling their resources. The VOs define the resources that are available (data, computers, applications, equipment etc.) aswell as protocols that allow applications to determine the suitability and accessibility of those resources. Sharing of data is defined by the relationships between the VOs that make up the grid, and the nature of these relationships is what distinguishes different data grids and the technology needed to implement them. Examples of data grids are those being set up at CERN to analyse the huge amounts of data generated by various high energy physics experiments [3].

Peer to Peer Networks. Peer to peer (P2P) networks [4] are one of the most researched technologies in computer science at the moment. They are designed for the sharing of various types of computer resources (content, CPU cycles etc) by direct exchange rather then requiring the mediation of a central server. Usually a P2P network focuses on sharing a particular type of resource. Use of peer to peer networks is motivated by their ability to "function, scale and self-organise" without the need of any central administration - each node in a peer to peer network is autonomous and equal to all the others - and the network grows in an ad-hoc fashion. Examples include SETI@Home [5] where computing resources are shared and file sharing networks like napster and gnutella [6].

Distributed Databases. Distributed databases [7] are usually used to integrate existing databases, thus allowing a single interface to be created for database operations. The interface is managed by a single entity and therefore distributed databases have a stable configuration with the data organised in a relational manner. By nature the data has a homogenous structure since this is enforced by the overall database schema.

2   Objective

Although there have been some efforts in the fields of computational chemistry and biochemistry to create comprehensive simulation databases [8] or databases of point calculations for specific fields [9] we feel there is a need for a general framework that allows easy and immediate sharing of molecular simulation data between collaborators and other researchers in the field.

Our main goal is to implement a method for sharing and searching the data generated by our molecular simulation program Adun [10], initially focusing on group and collaborative level sharing. This data can be divided into four categories - simulations (trajectory and energies), data sets, systems (collections of molecules and associated topology) and templates (blueprints for simulations). In addition all data has metadata associated with it e.g. keywords, textual descriptions, user annotations etc.

We want to maintain, as far as possible, the relationship between these data i.e. who generated what and how. This "data awareness" is vital for validating the usefulness of data and for enabling users to see what analysis has already been performed.

In addition to storing and maintaining this web of relationships we want access to the data to be transparent and pervasive. Display and searching of data must be integrated into the application and interaction with remote and locally stored data should proceed through the same mechanisms. In this way we remove all barriers to accessing the data and make it extremely easy for users to make their data available.

An important issue that arises in scientific data sharing is the need to share at different levels - from completely public, down to group and private levels. Hand in hand with this comes the need for different privileges for manipulating the data e.g. read, write, delete, move etc. Hence our solution must enable varying levels of data access and manipulation.

Since scientists work on numerous projects simulataneously, the ability to collect data related to these projects in user defined groups has a significant impact on productivity. To this end we require a way to group and store references to remote data, as found through previous searching, and to share these collections of references with collaborators.

In order to fulfill these goals we have implemented a peer to peer network based on PostgreSQL databases for storage and sharing of our data. This framework has many advantages that derive from both the free, powerful and robust nature of PostgreSQL and the inherent properties of a P2P network and we believe that it can be extended to other types of complex systems modeling techniques.

2.1   An Example

The utility of such a network is easily demonstrated by the following scenario. A computational biochemist wishes to calculate the binding free energy of GTP to a guanidine nucleotide binding protein e.g. Ras. They have previously found a Ras minimisation calculation on the network and have stored a reference to it in a personal project. Accessing the minimisation they extract the final structure which they save and add to their project. On saving a relationship is added to the structure, indicating where it was generated from, and to the minimisation data, indicating that data has been generated from it.

Next the scientist searches for GTP simulations and structures and finds a charge optimised configuration derived from a simulation using the ASEP (Averaged Solvent Electrostatic Potential) method [11]. Using the data knowledge capacity the researcher examines the simulation's initial options and finding them satisfactory adds the GTP system to their project.

Using the extracted systems the scientist runs a binding calculation using the linear interaction energy (LIE) [12] protocol, specifying that it be stored in a shared database on completion. Later another researcher working in collaboration checks the shared database for new developments. Finding the LIE simulation they access the data and check the computed result for the binding free energy.

3   Characteristics of Our Network

We are creating a network for content distribution which is by far the most common form of P2P network. These networks can be characterised by a number of properties as outlined in [13]

Scalability. This is how the performance of the network is influenced by the number of nodes in the system. In a scalable network a dramatic increase in the node population will have minimal impact on performance. Scalability is often determined by the number of nodes that can connect to a given node and the method used to query the network.

Performance. The time required for performing an operation e.g. addition, retrieval and querying of data.

Fairness. A fair network ensures that users offer and consume resources in a balanced manner. In order to do this many systems use reputation and resource trading mechanisms.

Resource Management. At a basic level content distribution systems must provide a means for publishing, searching and retrieving data. More advanced systems can provide metadata operations and relationship management.

Grouping. Grouping refers to the ability of the network to apply higher levels of organisation to the data it contains. Most networks allow only minimal grouping by file extension.

Security. The security features of a network can be analysed in three subcategories: Integrity and authenticity, privacy and confidentiality and avability and persistence.

Depending on the network certain properties are more desirable than others and hence influence the design decisions taken when implementing it. In turn these design decisions will feedback and affect other properties. Also some network characteristics are implied or expected by the type of data being shared and who is sharing it.

Our network is for sharing scientific data between groups of collaborators with the aim of providing a "data awareness" facility and powerful querying. Based on this we discuss its properties both implied and desired, in the context of the above categories.

3.1   Resource Management and Grouping

We are focusing initially on providing advanced data management and organisation features, so the resource management capabilities of our network are paramount. The data we are sharing is homogenous since it is application specific. This homogeneity greatly simplifies our task since many of the problems faced by other networks are due to heterogenous content. In addition the collaborative basis of our network and the scientific nature of the data should naturally lead to nodes containing similar data to be grouped together. Further layers of organisation will be added atop the network topology by both the data awareness of the content and by the ability of users to create groups of data based on their research interests.

3.2 Security

Certain security issues are a major concern for our implementation. In terms of privacy and confidentiality we require multiple access levels to the data so its use can be restricted to certain groups.

We ensure the integrity of our data in a number of ways. It is immutable, preventing it from being maliciously tampered with and the privileges to add, delete and update data are strictly controlled. Since we aim to store information about what and how each piece of data was generated this will enable the quality of the data to be judged by the users themselves.

Finally we expect the availability and persistence of data on our network to be aided by the nature of the users. It is reasonable to assume, based on experience, that scientific computers are shut down much less frequently than home computers. This coupled with the expectation that peers will also be collaborators, leads us to predict much less volatility than is normal in such networks.

3.3 Scalability

The nodes of our network are likely to be collaborators or researchers with similar interests and it has been shown that networks based on such interactions are "small-world" in nature - each node is only directly connected to a small number of other nodes [14]. In this case new nodes are likely to be distributed around already existing clusters or to form new ones and therefore will have minimal impact on performance and availability.

3.4 Performance

Although network performance is important, it takes a back seat to resource management and security in this initial implementation. The first iteration of this network will be strictly peer to peer so we can eliminate the need for query forwarding facilities thus maintaining simplicity and avoiding many complicated performance related issues. Therefore performance will mostly depend on download time from remote hosts, especially when dealing with gigabyte sized trajectory data, and it is in this area that we will concentrate most of our performance enhancing efforts.

However we realise that as the amount of data grows query execution time will also become a factor and must be taken into account so the network can adapt to handle it in future iterations. Here we can take advantage of both the data awareness of the content and the existence of projects.

3.5   Fairness

This property has the least weight in our network. Here we expect the collaborative nature of the network itself to introduce fairness as a basic characteristic. In addition we benefit from the fact that consumption of our data automatically implies production of new data.

4   Design Issues

There are a number of general design issues that one is faced with when implementing a network [15]. These are listed below along with the properties they most influence.

Network Structure. (Scalability, Performance, Privacy, Availability) There are two extremes of network structure. On one hand we have structured networks where the topology is tightly controlled and on the other are unstructured networks where the topology is ad-hoc and indeterministic.

Content Replication. (Scalability, Performance, Availability) Often peer to peer systems rely on the replication of data on many nodes to improve availability and performance. Many different schemes exist, from simple passive replication where copying occurs naturally as users copy content, to active replication where the network itself copies and moves data based on usage patterns.

Access Controls, Deniability and Secure Storage. (Security) These three issues relate to the overall security properties of the network.

Accountability and Reputation. (Fairness, Privacy) In many file sharing systems it is necessary to stimulate cooperative behaviour to prevent users only consuming resources and not providing anything. Many mechanisms have been developed to cope with this problem such as reputation systems where users are given preference based on their degree of participation.

Content Storage and Management. (Resource Management) This refers to the handling of a collection of related tasks such as searching, storage and bandwidth management.

How each of these issues is dealt with is determined in part by the desired and implied properties of the network and in part by design constraints e.g. resources, time, manpower etc. The biggest constraints on our design are our wish for the network to be relatively easy to implement and for it to be an optional extension to our program. To achieve the first objective we must exploit existing free technologies. Since we are focused on data management the use of an advanced relational database to store the data is invaluable, enabling features such as transactions and concurrency control, querying, and robust data recovery and backup. By utilizing a relational database as the backbone of our system we not only gain data storage facilities but also server, user management and security features thus covering a number of the design issues at once. In this case we have chosen PostgreSQL and below we detail how it is used and how it affects the decisions made in various areas.

The second constraint requires that we integrate storage on the network with the default file based database used by Adun. The same information must be stored in both and movement between SQL local databases must be transparent.

4.1   Content Storage and Management

Most technologies use simple file based storage which has many limitations since it effectively ignores metadata. This has lead some researchers e.g. [16] to begin using relational databases to store content, thereby giving the data a rich structure aswell as allowing powerful querying, which is the approach used here.

An advantage of PostgreSQL in this area is that it provides two special mechanisms for handling large amounts of binary data, which is the bulk of what we will be storing. The first is TOAST (The Oversized-Attribute Storage Technique) which allows storage of binary data up to 1GB in a single field. The second method is known as "large objects". Although not as simple to use as TOAST it has other advantages including the ability to store data up to 2GB in size and to allow streaming reads and writes of the data.

The database server also enables fine grained control of the number of simultaneous connections allowed and resource usage, allowing users to fine tune it to suit their needs.

4.2   Network Structure

For this initial version we adopt a purely unstructured topology. Connection is left to users of the software who must advertise their servers to other users and queries will be strictly peer to peer with no query forwarding. This is not only a result of our simplicity constraint but is also caused by using the database server exclusively to accept connections. Since the list of known peers of each node is maintained by the program the database server knows nothing about the rest of the network and cannot pass queries to other nodes.

4.3   Access Controls and Secure Storage

PostgreSQL also handles most of our security issues. Data integrity is one of the main features of a database. Transactions enable blocks of SQL statements to be executed as all or nothing events. If any error happens during the block the transaction can be "rolled back" and the database restored to its initial state. Transactions are closely related to concurrency which isolates users who simultaneously operate on the same data. PostgreSQL has the advantage of using multi-version concurrency control which enables reading and writing to occur in a non-blocking manner. The databases privilege system also limits who can manipulate the data.

Access control is handled both by the server, which can be configured to only accept connections from certain clients, address ranges or users as well as to employ various password based access schemes, and by the database itself. In the case of the database we will utilise PostgreSQL's schema facility to enable data to be restricted to certain users.

Due to the nature of our network strong anonymity features are not possible. The network is currently formed by people who must explicitly notify each other when they wish to share data. In addition the ability to use a given database for storage is determined exclusively by the owner of that database.

4.4 Content Replication and Accountability

Content replication is of a passive nature - this is both due to the simplicity of our network and also to address efficency e.g. avoid replicating large amounts of data. However to enhance performance local caches of data will be created on each node allowing users quick access to commonly used data without the need to copy it explicitly.

As mentioned previously we do not foresee problems related to uncooperative users that are endemic to some other networks. Hence no accountability or reputation system is used.

5 Architecture and Implementation

In this section we discuss in detail the architecture and design of the network addressing key implementation features. An abstract overview of the structure is shown in Fig. 1.

5.1 The Data

Each of the four data types, systems, simulations, templates and data sets, is represented by a class in Adun and what we store are serialised versions of program objects along with related metadata e.g. keywords, annotations, descriptions etc.

When a data object is instantiated a globally unique ID is generated for it. These ID's are used as the primary keys in our database. The creator and creation time of each object is recorded along with references to the objects it is derived from. These references take the form of a unique ID, class type, and current location. Issues related to updating output references are discussed below. The version of the program used to create the object is also saved.

5.2 The Database

Schemas. To enable different levels of data sharing we utilise the PostgreSQL concept of schemas. Each schema can be thought of as a space in the database requiring certain permissions to enter.

Another property of schemas that we exploit is that tables in separate schemas can have the same name. This allows us to define a "template schema" - a collection of tables and the relationships between them. The template is a blueprint with particular instances of it having different names.

Our implementation utilises two template schemas - Adun Objects (see Fig. 2) which stores data and Adun Projects for holding project information. Each database can contain multiple instances of these templates e.g. PublicAdunObjects, GroupAdunObjects, PrivateAdunProjects (the names are up to the user) etc. In addition each database will also contain a schema called ExternalDataReferences, used to enable data awareness.

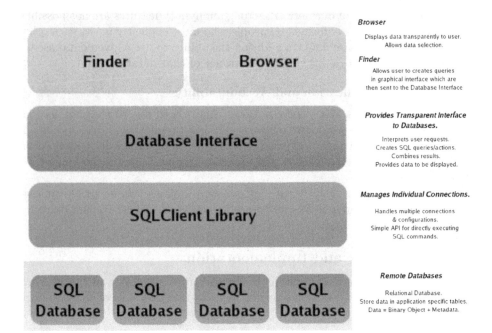

Fig. 1. Abstract overview of our implementation. At the bottom we have a number of remote SQL databases which form the nodes of the network. Communication with the nodes is handled by the SQLClient library. The top two layers are part of our program Adun.

Adun Objects. The most important part of creating a database is defining its structure e.g. the tables and the relationships between them. The structure of an Adun Objects schema is summarised by an entity relationship diagram (see Fig. 2). The main entities in the diagram correspond to each of the data types while the other entities define the "many to many" relationships between the data. In addition there is a table for keywords and relationship tables that associate keywords to data and vice versa.

As can be seen from the ERD the relationships between the data generated by Adun is complicated - excluding templates every type of data can be derived from and in turn can create every other type.

ExternalDataReferences. ExternalDataReferences allows storage of references to data outside a schema. It consists of a table for each data type, the entries of which contain the unique id and name of an object along with its location i.e. schema.database.host. Use of this schema requires extra relationship tables to be added to Adun Objects. These tables have the same structure as the intra-schema ones but relate data in the schema to data in ExternalDataReferences e.g. along with the table SystemDataSets there is also ExternalSystemDataSets etc.

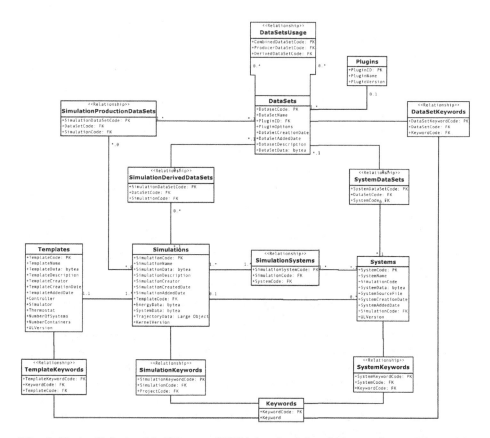

Fig. 2. Entity Relationship Diagram (ERD) for the Adun Objects schema. The multiplicity and cardinality of the main relationships are indicated. The fields storing binary data are labled - "bytea" indicates storage using TOAST while "large object" indicates the large objects facilitate is used. PK and FK stand for Primary Key and Foreign Key respectively.

Adun Projects. Each project schema consists of a table containing the project name and ID plus tables handling the many to many relationships between projects and the Adun data types. Entries in a relationship table consist of a project id and a foreign key attached to the corresponding table in ExternalDataReferences.

5.3   Database Interface

The database interface layer consists of three classes, one for handling SQL Databases (ULSQLDatabaseBackend), one for handling the file system database (ULFileSystemDatabaseBackend) and the overall database interface (ULDatabaseInterface) respectively. Each connection to a SQL database as a given user is represented by an instance of the ULSQLDatabaseBackend class. This class utilises the SQLClient library to connect to, and execute SQL statements on, the database in question. The database interface instance creates and manages all client connections keeping track of which connections are available, notifying observers when connections come online or are lost. It also collates the results of queries and resolves requests for remote references.

SQL specific functionality is encapsulated by ULSQLDatabaseBackend which adopts the same interface as ULFileSystemDatabaseBackend. In this way the program is completely insulated from how and where the data is stored, enabling us to distribute the SQL Backend functionality as a extension plugin for the program.

Caching. In order to speed up access to commonly used data the middle layer creates caches for each database it connects to. Temporary copies of downloaded objects are stored in the cache of the related database which are searched first when a retrieval request is made. Each cache is controlled by a cache manager who maintains a user defined size limit on the data stored in the cache by removing the least used data present.

Streaming. The size of system, template and data set data is usually of the order of 10 MB's. Simulation data is made up of energy data, which is also of the order of 10 MB's, and trajectory data which can reach gigabytes in size. Obviously transfer of trajectory data is the main performance concern. A trajectory consists of a sequence of configurations of a set of atoms and it is usual to operate on one configuration at a time. By storing trajectories using PostgreSQL's "large objects" facility we can access their data using a stream and read configurations on demand instead of downloading them all at the same time, resulting in greatly enchanced performance.

Metadata Updates. Metadata is stored in both the archived object and in the columns of the relevant database table. Since retrieving and updating an object on a metadata change would put a huge strain on resources, metadata updates are only performed on the table columns. However this in turn means that the archived metadata will become out of synch over time. To overcome this when an object is retrieved its metadata is synched to the values currently in the table by the ULSQLDatabaseBackend instance and the archived values are only changed whenever the object is copied or moved.

Data Awareness. Data awareness comes in two forms - knowledge about who the data was derived from and knowledge about what has been derived from it, called here input and output knowledge

The main problem in maintaining both types of knowledge derives from the necessity to limit the write privileges of unknown users of the data. This manifests itself both when initially creating references between data and when moving, removing or copying it.

Creation of new data implies creation of at least one input and one output reference. Input references are easier to handle. In order to be able to add data to a database schema the user must have the SQL INSERT privilege and hence can also add internal schema references. In addition the ExternalDataReferences schema is writable to any one who already has write privileges on a schema so external references can also be added.

Output references are more problematic since they require the user to have write permissions to where the data is stored. This can never be true for data from a schema where the user has read only access and output referencing is disabled in these cases i.e. once data is accessible to read only users you cannot track everything that has been derived from it.

When a user wishes to retrieve the data referred to by a remote reference the database interface searches for an available client who can access the given database and schema and retrieves it from there or displays a message it if cannot obtain the data.

Moving and Duplication. When data is copied all references are copied with it. The main task when copying is to ensure that when references are created in the new database that local ones are identified and dealt with accordingly. However knowledge about data derived from the copies is not available to the original data.

Moving is more restrictive. In addition to being able to remove data from a schema and add it to another the user must have the UPDATE privilege, and access to, the location of all its input data and output data. Although this is quite restrictive it naturally ensures that heavily used data remains available i.e. makes it difficult for someone to move it from the public domain for instance.

However nothing can be done when a user with the necessary permissions wishes to removes data. References to deleted data are kept but attempting to access it results in a "Data not present" message being displayed.

5.4   User Interface Layer

At the user level data is displayed by a browser in the main Adun window (see Fig. 3). It is grouped by client connection, schema and then data type. The contents of the RemoteAdunReferences schema are not included. The browser enables cut, copy, paste and remove operations on selected data and also shows the known antecedents and descendants of each object. The class managing the view receives updates from ULDatabaseInterface when data is added or removed from a database, or when a client becomes disconnected or available, and updates the display accordingly. Formation of queries is handled by a finder tool with the results being displayed by the browser.

6   Practical Details

Probably the main disadvantage of this implementation is that most of the functionality is determined by the database which requires the user to have at least some knowledge of PostgreSQL internals. What follows is a quick overview of some of the practical issues that must be tackled. Note that as our program evolves a number of the issues addressed here will become obsolete.

Fig. 3. Screenshot of the main Adun window. Users access and manipulate data through the browser. Metadata on each object can be retrieved using the properties command and the data itself can be edited (cut, copy, paste etc), displayed, loaded for analysis or used to create simulations.

6.1 Database Creation and Roles

The database itself is created via an SQL script that will be distributed with the extension package. This script initially creates an objects schema called PublicAdunObjects, a project schema called PrivateProjects and the External-DataReferences schema. The owner of the database and created tables will be the user the script was run as.

PostgreSQL supports a number of privileges -

1. Database Level
   (a) CREATE - Add new schemas
2. Database Level
   (a) USAGE - See whats a schema contains.
   (b) INSERT - Add data
   (c) UPDATE - Change metadata
   (d) DELETE - Remove entries
   (e) SELECT - Perform queries and retrieve data

Initially a role called *Adun* will be created. This role has login rights to the created database, and has select and usage privileges on the PublicAdunObjects schema, but no write permissions. Hence you cannot add to the database using this role and output reference tracking is disabled.

The user who ran the script has all permissions on all objects. The script also removes the public schema which is created by default when a database is created. This is because every role has CREATE privileges on it by default which is not desirable.

At the moment creation of other roles and granting of privileges is only possible from the database interface.

6.2   Security

The security procedure for accessing data is very rigorous and highly customisable. There are four security levels that must be passed:

First Level - Host Connections. The first access requirement is that the database port must be open and accepting requests. Of course any firewall can be extensively configured to limit who can connect.

Second Level - Connecting to the Server. A computer can have a number of IP addresses assigned to it however the PostgreSQL server will only listen for connections on address specified in the listen_address section of the *postgresql.conf*.

Third Level - Connect to the Database. In order to connect to any database with name *User* there must be an entry in the PostgreSQL *pg_hba.conf* file. A default entry for user *Adun* must be entered to allow basic connections e.g. the following two are recommended

1. host sameuser Adun (ip options) trust
2. host samerole Adun (ip options) md5

The first entry allows anyone connecting as user *Adun* to access the Adun database (assuming they've passed the previous two levels of security and meet the restrictions set by ip options). As described previously the role Adun is defined to only allow reading of the public schemas hence we use trust as the authentification method.

The second entry enables users to connect with roles that inherit from the *Adun* role. Depending on the role the person connecting can have a variety of privileges so md5 security authentification is used for these connections i.e. password required.

Fourth Level - Role Privileges. Once someone passes through the above three levels they have access to the database. The fourth level of security is determined by the role they connected as and the privileges assigned to it.

6.3   Installing and Connecting

To enable the data sharing functionality an extension bundle must be built and installed. Building the extension requires the SQLClient library and the PostgreSQL C interface library to be installed. Once the extension is installed data sharing is enabled the next time Adun is started. New databases can be added through the Adun interface which requires a username, database name and host name or address. When a node is added its data becomes available for manipulation based on the privileges of the user you are connected as.

6.4   Performance

Although network performance is not a main concern at the moment we performed preliminary tests using a network including computers from both our own group in Barcelona and other hosts located in London. We found that download and upload times are, as expected, heavily dependant on the quality of the connection between the two nodes. However use of streaming for reading data and threading for addition makes a noticeable difference. It should be noted that the difference is not due to any increase in the download rate. Rather it is a result of reducing the percieved time tasks take by running them in detached background threads or deferring retrieval of data until it is requried. In depth testing of query performance was not possible at the time of writing since no database included enough data to make querying time a noticeable problem.

6.5   Other Configuration Issues

The basic configuration of PostgreSQL is usually acceptable for most users needs. However parameters concerning the maximum number of connections and the general resource usage of the database can be fine tuned if necessary through the postgresql.conf file. Detailed information on what can be configured can be found in the PostgreSQL manual.

7   Conclusions

We have implemented a method for sharing the data generated by our molecular simulation program Adun through the use of a peer to peer network built on PostgreSQL databases. The initial implementation concentrates on providing high level data management facilities by utilising the inherent features of relational databases. Data is stored along with metadata, such as keywords and descriptions, which can be used in queries. We also employ the concept of projects, which allow users to save references to remote data in groups, and enable "data awareness" whereby users can see what generated and what has been generated from a particular data object.

The network also includes some basic performance enhancing features such as streaming of large trajectory data, caching of commonly used objects and threaded addition of data. It has been tested linking computers across a local network and computers at remote sites and it was found that performance in terms of retrieval time is determined by the bandwidth of the connection between the two nodes in question.

This network is only an initial implementation and it can be improved in many areas. Additions could include a web based directory which provides a list of hosts that users can connect to along with brief descriptions of the type of data they contain. Also query forwarding is likely to be implemented to aid data discovery e.g. when searching, each node propagates the query to nodes it is connected to. This flood based querying normally has major scalability problems since it cannot know where a file of a given type is but it also cannot propagate a query over the entire network. However much research has been done on this issue [17,18] and in our case we can exploit both the intrinsic collaborative topology of the network [14] as well as the existence of projects and data awareness.

Addition of a central server to aid in retrieving data is also a possibility. An advantage of this method is that by giving the central server the privilege to insert references in all nodes we can circumvent the output referencing problem i.e. the program can use the central server to add output references when the user does not have the correct privileges themselves.

Future versions will also concentrate on improving query performance by taking advantage of PostgreSQL's many facilities for query optimisation as well as providing efficient full text searching of the text description metadata field. Maintenence and interaction with the databases will be simplified by the implementation of administration tools allowing the creation of new schemas and users along with the managing of roles and privileges from Adun.

In conclusion we believe the method outlined here is relatively simple to implement and provides an ideal way for users of complex systems applications to share and collaborate on the data they produce.

Acknowledgments

This work was supported in part by Spanish MCYT grant BQU2003-04448. MAJ holds a PhD fellowship from the Generalitat de Catalunya.

References

1. NIH: NIH Data Sharing Policy and Implementation Guidance. (2003)
2. Venugopal, S., Buyya, R., Ramamohanarao, K.: A taxonomy of data grids for distributed data sharing, management, and processing. ACM Comput. Surv. **38**(1) (2006) 1–53
3. Lebrun, P.: The large hadron collider, a megascience project. In: Proceedings of the 38th INFN Eloisatron Project Workshop on Superconducting Materials for High Energy Colliders. (1999)
4. Oram, A.: Peer to Peer: Harnessing the Power of Disruptive Technology. O'Reilly and Associates Inc., Sebastopal, CA (2001)
5. Anderson, D., Cobb, J., Korpela, E., Lebofsky, M., Werthimer, D.: Seti@home: An experiment in public-resource computing. Comm. ACM **45**(11) (2002) 56–61

6. Gnutella: The gnutella website. (http://gnutella.wego.com)
7. Mark, W.L.L., Roussopoulos, N.: Interoperability of mulitple autonomous databases. ACM Comput. Surv. **22**(3) (1990) 267–293
8. Model: A database of molecular dynamics simulations. (http://mmb.pcb.ub.es/MODEL)
9. Giese, T.J., Gregersena, B.A., Liua, Y., Nama, K., Mayaana, E., Mosera, A., Rangea, K., Nieto Fazaa, O., Silva Lopez, C., Rodriguez de Lerab, A., Schaftenaare, G., Lopez, X., Leea, T.S., Karypisc, G., York, D.M.: QCRNA 1.0: A database of quantum calculations for RNA catalysis. Journal of Molecular Graphics and Modelling1 **in press** (2006)
10. Johnston, M., Fdez. Galván, I., Villà-Freixa, J.: Framework based design of a new all-purpose molecular simulation application: The Adun Simulator. J. Comput. Chem. **26** (2005) 1647–1659
11. Sánchez, M.L., Aguilar, M.A., Olivares del Valle, F.J.: Study of solvent effects by means of averaged solvent electrostatic potentials obtained from molecular dynamics data. J. Comput. Chem. **18** (1997) 313
12. Aqvist, J., Marelius, J.: The Linear Interation energy Method for Predictiong Ligand Binding Free Energies. Combinatorial Chemistry and High Throughput Screening **4** (2001) 613–626
13. Androutsellis-Theotokis, S., Spinellis, D.: A survey of peer-to-peer content distribution technologies. ACM Comput. Surv. **36**(4) (2004) 335–371
14. Chirita, P.A., Damian, A., Nejdl, W., Siberski, W.: Search strategies for scientific collaboration networks. In: P2PIR'05: Proceedings of the 2005 ACM workshop on Information retrieval in peer-to-peer networks, New York, NY, USA, ACM Press (2005) 33–40
15. Rodrigues, R., Liskov, B., Shrira, L.: The design of a robust peer-to-peer system. In: EW10: Proceedings of the 10th workshop on ACM SIGOPS European workshop: beyond the PC, New York, NY, USA, ACM Press (2002) 117–124
16. Ooi, B.C., Shu, Y., Tan, K.L.: Relational data sharing in peer-based data management systems. SIGMOD Rec. **32**(3) (2003) 59–64
17. Chawathe, Y., Ratnasamy, S., Breslau, L., Lanham, N., Shenker, S.: Making gnutella-like p2p systems scalable. In: SIGCOMM '03: Proceedings of the 2003 conference on Applications, technologies, architectures, and protocols for computer communications, New York, NY, USA, ACM Press (2003) 407–418
18. Nejdl, W., Siberski, W., Sintek, M.: Design issues and challenges for rdf- and schema-based peer-to-peer systems. SIGMOD Rec. **32**(3) (2003) 41–46

Accessing Bio-databases with OGSA-DAI - A Performance Analysis

Samatha Kottha[1], Kumar Abhinav[1,2],
Ralph Müller-Pfefferkorn[1], and Hartmut Mix[1]

[1] Center for Information Services and High Performance Computing,
TU Dresden, Germany
http://tudresden.de/
die_tu_dresden/zentrale_einrichtungen/zih/forschung/grid_computing
[2] Vellore Institute of Technology, India
{samatha.kottha, ralph.mueller-pfefferkorn, hartmut.mix}@tu-dresden.de,
abhinav.vit@gmail.com

Abstract. Open Grid Service Architecture - Data Access and Integration (OGSA-DAI) is a middleware which aims to provide a unique interface to heterogeneous database management systems and to special type of files like SwissProt files. It could become a vital tool for data integration in life sciences since the data is produced by different sources and residing in different data management systems. With it, users will have more flexibility in accessing the data than using static interfaces of Web Services.

OGSA-DAI was tested to determine in which way it could be used efficiently in a Grid application called RNAi screening. It was evaluated in accessing data from bio-databases using the queries that a potential user of RNAi screening would execute. The observations show that OGSA-DAI has some considerable overhead compared to a JDBC connection but provides additional features like security which in turn are very important for distributed processing in life sciences.

Keywords: OGSA-DAI, Grid computing, performance analysis, bio databases.

1 Introduction

With the latest technology of high throughput experiments and technology, the amount of data produced in life sciences is increasing exponentially. The data coming from different experiments are stored in different formats and maintained by researchers around the world. During the development of an application which accesses the data from different resources or an application, which requires a huge amount of computing resources like a typical Grid application, accessing the data in a fast and secure way becomes a very critical issue. In case of first application type, different data providers have to provide an application interface to their data. A few are providing Web Services to retrieve data from their databases but

W. Dubitzky et al. (Eds.): GCCB 2006, LNBI 4360, pp. 141–156, 2007.

these webservices are not dynamic or flexible enough to adopt the ever changing user needs. The application developer is forced to create and maintain his/her private databases of the publicly available data, which is a overlapping and redundant work. If an application requires huge amount of computing and storage resources which could not met by local cluster, the only choice is to use grid computing. In such a case, distributing and maintaining the heterogeneous data and retrieving it in secure mode becomes very important. As the collaboration and cross-disciplinary research become the new mantra in life sciences, the need for tools to access and integrate data coming from different data resources with ease of use is increasing. OGSA-DAI could fill this role comfortably.

Open Grid Service Architecture - Data Access and Integration (OGSA-DAI) [1] was conceived by the e-Science project initiative in UK. The aim of OGSA-DAI is to provide a unique interface to access and integrate data from separate sources via the grid. It provides a transparent way of querying, updating, transforming, and delivering data via web services. It hides the complexity of heterogeneous databases storage and location from the user. For example, the user will be able to access a MySQL database or an Oracle database with the same interface.

RNAi screening is one out of four bioinformatic subprojects in the German MediGRID[1] project that kicked off in fall of 2005. It has three major components: threading pipeline, Structural Classification of Protein-Protein Interfaces (SCOPPI) [2] database, and docking algorithm. Upon completion, it will provide a portal interface for users to enter the sequence of the interesting protein in the FASTA format to predict the structure if its not yet known, to find the interaction partners, and to bind. Behind the portal is a threading pipeline [3],[4],[5] which predicts the structure of a protein, returns the 3-D coordinate file of the predicted structure and the PDB IDs of the proteins that are having a structure similar to that of the predicted structure of a requested protein. The functional domains of these proteins would be determined from Structural Classification of Proteins (SCOP) [6] and their interaction partners using SCOPPI. SCOPPI is a comprehensive database that classifies and annotates domain interactions derived from all known protein structures. At present it considers only SCOP domain definitions and provides interface characteristics like type and positions of interacting amino acids, conservation, interface size, type of interaction, GO terms, GOPubmed [7] links, and screenshots of interfaces and participating domains etc.

After finding the interaction partners of functional domains, users will be given the option of choosing the interactions. The user can dock the proteins that are interesting to him or her using a rigid body docking algorithm BDOCK [8] at granular angle of $2°$. The threading and docking are very computing intensive tasks and require a huge amount of resources. The preparation of data for the SCOPPI database also requires a large number of computing resources but only from time to time. It is recalculated only when a new version of SCOP is released, which is twice in a year utmost.

---

[1] http://www.medigrid.de/

In its workflow, RNAi application accesses several databases which are located at different geographical locations and use different database management systems. For example, the Macromolecular Structural Database (MSD)[2] is an Oracle database whereas SCOP and SCOPPI are MySQL databases. It also accesses flat files like SwissProt and PDB files. The screenshots of interfaces are at present stored in a file system and their path is stored in the database.

Usually, the database and file servers are kept behind firewalls. To make them accessible to an application, either the servers need to be placed outside the firewall or the ports need to be opened for all the machines in the Grid where the application might run. This poses a severe security threat. One possibility to circumvent this problem is to use OGSA-DAI to deploy the databases and to use GridFTP[3] or Grid Reliable File Transfer (GridRFT)[4] to move files around in Grid. In both cases the database and file servers just need to open ports for limited number of machines where OGSA-DAI and GridFTP servers are running.

Another possibility would be to use the Storage Resource Broker (SRB) [9], which was also tested. However, the current limitations of SRB in accessing database management systems (only Oracle databases currently) prevent an efficient use.

Since RNAi screening application is still in development stage, its not yet known how many users will be using the RNAi screening application simultaneously and how intensively. However, it requires to access several MySQL databases like SCOP, SCOPPI and Oracle database MSD and Protein Data Bank (PDB)[5], Protein Quaternary Structure (PQS)[6], and image files. The user query results will be less than 1 MB unless they are retrieving a large number of image files. Allmost all of the data centers that are part of German Grid are connected by a 10 GBit/s network. Encryption of the data during the transfer is not a compulsory feature but transport level security is desired.

2 OGSA-DAI

OGSA-DAI is an extensive and extendable framework for accessing data in applications. It supports relational databases like Oracle, DB2, SQL server, MySQL, and Postgres, and XML databases like xindice and eXist, and files like CSV, EMBL, and SwissProt files. Databases are deployed as data service resources, which contain all the information about the databases like their physical location and ports, the JDBC drivers that are required to access the databases, and the user access rights. A data service exposes the data service resource in a web container, which could be a globus container[7] or Apache Tomcat[8] server.

---

[2] http://www.ebi.ac.uk/msd/
[3] http://www.globus.org/toolkit/docs/4.0/data/gridftp/
[4] http://www.globus.org/toolkit/docs/4.0/data/rft/
[5] http://pdbbeta.rcsb.org/pdb/Welcome.do
[6] http://pqs.ebi.ac.uk/
[7] http://www.globus.org/toolkit/docs/4.0/
[8] http://tomcat.apache.org/

Basically, OGSA-DAI architecture can be divided into three layers: data layer, which contain the databases, business layer, which is core of OGSA-DAI and contains a data service resource for each database/file system, and the presentation layer, which encapsulates the functionality required to expose data service resources using web service interfaces. In this case study, the OGSA-DAI server refers to the machine which contains the business and presentation layers, the local and remote databases refers to the data layer.

In OGSA-DAI, activities are operations like data resource manipulations (e.g. SQL queries), data transformations (e.g. XSL-T transformations) or data delivery operations (e.g. GridFTP), that data service resources perform on behalf of a client. Factory is a service to create a data service instance to access a specific data resource. The client sends a XML document called perform document, which specifies the activities to be executed on the data service. On return it receives a response document, which contains the execution status of the perform document as well as query data.

OGSA-DAI is available in two flavors OGSA-DAI WSRF (Web Services Resource Framework), which is compatible with the Globus Toolkit implementation of WSRF and OGSA-DAI WS-I (Web Services Inter-operability), which is comapatable the UK OMII's implementation of WS-I[9] and could be deployed on Tomcat server.In this study, the OGSA-DAI WSRF2.2 performance is evaluated.

OGSA-DAI is providing a number of activities[10]. The most important activities that were used for the evaluation can be found in Table 1.

3   Methods

The tests were divided into two parts: loading data and retrieving data. Loading of files from the client machine and loading temporary data like results of a query into another table were evaluated. OGSA-DAI provides a feature called SQLBulkLoad (sqlBulkLoadRowset activity), which is required when the users need to join tables from multiple data resources. Using this feature one can load a query result into a table in another database. For retrieving data, the main deliver options provided by OGSA-DAI (shown in Table 1) are evaluated. All the tests conducted using the databases on a remote machine (a database server without OGSA-DAI) located at Zuse Institute Berlin (ZIB)[11] as well as on a local machine (the database server and OGSA-DAI residing on the same machine), which would be the landscape of RNAi project in production. The performance of OGSA-DAI is measured either without any security or with full (transport level and message level) security. OGSA-DAI provides transport and message level security only from OGSA-DAI server to the client but does not provide any security for the connection between OGSA-DAI server and database server on a remote machine. So, when measuring the performance of OGSA-DAI with

---

[9] http://www.omii.ac.uk

[10] http://www.ogsadai.org.uk/documentation/ogsadai-wsrf-2.2/doc/interaction/Activities.html

[11] http://www.zib.de/index.en.html

Table 1. The OGSA-DAI activities used for performance evaluation

Type	Activity	Description
Relational	sqlQueryStatement	Run an SQL query statement.
	sqlUpdateStatement	Run an SQL update statement.
	sqlBulkLoadRowSet	Load query results into a table
Transformation	xslTransform	Transform data using an XSLT.
		Used to transform the results to csv format
Deliver	deliverFromURL	Retrieve data from a URL.
		Used to read the XSL file required for transformation.
	deliverToURL	Deliver data to a URL. It can use HTTP, HTTPS, or an FTP.
	inputStream	Receive data through a service's data transport operations.
	outputStream	Deliver data through a service's data transport operations.

full security on a remote machine, the security between the database server and the OGSA-DAI server was ensured with secure shell tunneling.

On average, RNAi screening user query results will be smaller than the query results in this case study. The purpose of this study is not only just to know what performance the RNAi screening user will experience in future but also to know, how well OGSA-DAI handles large amount of data. Even though this study had RNAi screening application as context, the results could be true for any other application which loads or retrieves large amount of data to and from the database.

SCOPPI provides the images of interacting domains and the interface. At present, these images are stored in a file system and their file location address is stored in a MySQL database. Thus, the performance of OGSA-DAI was evaluated by storing the images as BLOBs in a table of the SCOPPI database. This would make the maintainability of several thousands of images easier in Grid environment using the database authorization features to restrict the access of individual users but large number of user queries could bring the database server to halt.

In this study, only accessing databases is evaluated. Threading and SCOPPI execute several queries on databases but we chose only one query for each type of category. Typical user queries were changed to increase the size of the data that were retrieved from the database to create a more stressfull situation.

For each query, the time from the connection or service initialization to connection to service termination was measured. This includes the query execution time and getting the query resultset to the client when retrieving data but doesn't include display of results. JProfiler[12] was used for profiling. Five readings were taken for each measurement and the average is calculated. For each measurement the database buffers were flushed before and globus was restarted.

Various data loading and retrieving methods of OGSA-DAI were measured and compared with direct JDBC connection. A typical RNAi user is not allowed to update, delete or insert data from the databases MSD, SCOP, and SCOPPI, but however he/she can get his/her own MySQL database upon request to write temporary tables.

4   Results

4.1   Hardware and Software Configuration

The hardware and software used for evaluation are listed in Table 2 and the test landscape is shown in figure 1. The landscape comprises a typical OGSA-DAI installation as would be used in a real Grid environment with local databases (deployed on the OGSA-DAI server) and remote databases. The OGSA-DAI server and the remote database servers are behind firewalls. The client and OGSA-DAI server are in the same LAN whereas remote machine is connected over WAN.

Table 2. Hardware and Software Configuration used for the OGSA-DAI evaluation

	Client	Local Server	Remote Server
CPUs	1 x AMD Athlon 64 3800+ @2.4GHz	4 x Intel Itanium 2 IA -64 @1.4GHz	4 x Intel Xeon @2.2GHz
Memory	2 GB	8 GB	4 GB
OS	Suse 10.0 64-bit	Suse 9.3 64-bit	Red Hat 4.3
Java	1.4.2	1.4.2	1.5.0
MySQL	Client 4.1.13	Server 4.0.18	Server 4.1.12
Oracle	Client 10 g	Server 10.1.0.3	None
Globus	Client 4.0.2	4.0.2	None
OGSA-DAI	Client 2.2	WSRF 2.2	None

All the tests were conducted using the databases on local as well as remote machines except the test which retrieves the data from oracle database as there was no Oracle installation available on remote machine. This gives us rough estimation of the overhead caused by accessing data over WAN. The datatypes that are loaded and retrieved include all types of data like texts, floats, and BLOBs. The protein sequences and the interaction images are stored as BLOBS in the databases. The queries used for the evaluation are available upon request.

---

[12] www.ej-technologies.com

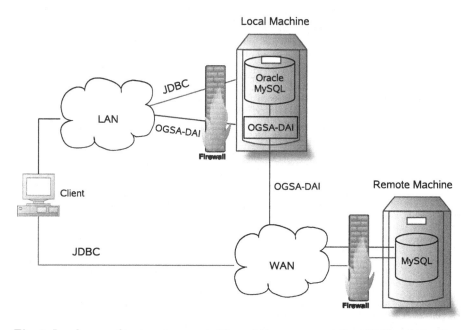

Fig. 1. Landscape of test environment. The red lines represent the JDBC and blue lines represent OGSA-DAI connections. The network bandwidth to local machine is ca. 9 MB/sec whereas to remote machine is ca. 4 MB/sec. The JDBC client could transfer data at 6 MB/sec at most.

4.2 Loading Data

In the category of loading data three scenarios were tested:

1. **Loading a file from the client into local and remote database:** None of RNAi users will be allowed to load or modify the data in the databases (SCOP, SCOPPI, and MSD) but some will get their private MySQL database upon request to load their own data. Size of the loading data is a hard guess but it could be anywhere between couple of thousand records to a million. Thus, in the tests, CSV files with 1000 to over 1 million records were loaded from the client to the local and remote MySQL databases using JDBC and OGSA-DAI. The record size was 101 Bytes. Using JDBC the file could be directly uploaded into a table in database as the SQL queries are executed by the client. However, in case of OGSA-DAI, the client transfers the SQL queries as plain strings and the queries are executed by the OGSA-DAI server. So, to load a file using OGSA-DAI the file has to be located on the OGSA-DAI server. This has been accomplished by transferring the file using Grid secure copy (gsiscp) to the OGSA-DAI server and then loaded. The performances of JDBC and OGSA-DAI with local and remote databases are shown in figure 2.

   In case of JDBC, the connection to the database took roughly 400 milliseconds whereas initialization of an OGSA-DAI service took on average 5 secs.

Loading Data To A MySQL Table

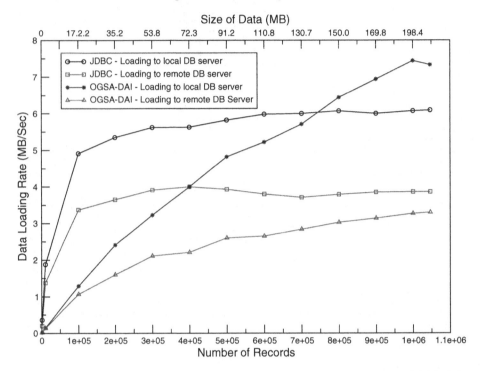

Fig. 2. Loading data from a file into a table using JDBC connection and OGSA-DAI connection. In case of OGSA-DAI, the file is transfered to the OGSA-DAI server using Grid secure copy. The records size is 101 bytes.

In case of OGSA-DAI, preparing the perform document and SOAP calls sent to OGSA-DAI server are additionally expensive. That is the reason the JDBC is doing very well when loading small files. But these expenses are compensated by the file transfer using gsiscp, which was able to transfer the file from client to the OGSA-DAI server at ca. 9 MB/sec rate. Whereas the JDBC client was able to transfer the file to local database server with a rate of utmost 6MB/sec. That is why OGSA-DAI is performing better than direct JDBC when loading large files into the database that is located on the same machine like OGSA-DAI server. However, when large files are loaded to a remote database, the transfer rate over WAN is on average ca 4 MB/sec, which is less than both JDBC and gsiscp transfer rates. So, OGSA-DAI is not able to take advantage of faster file transfer to outdo the JDBC.

2. **Loading images from the client into local and remote database:** At present, the protein-protein interaction images are stored in a file system and their logical names are stored in the database. Each image is roughly 60 KB of size. A new MySQL table is created with the protein-protein interaction

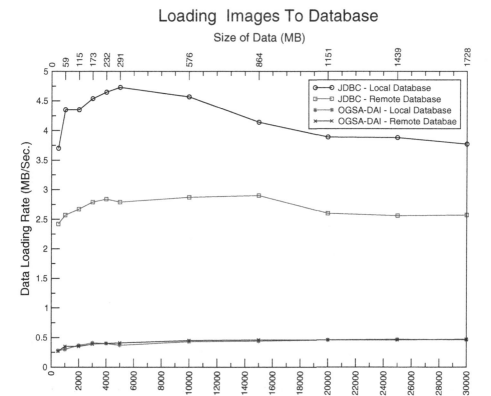

Fig. 3. Loading images into a table as BLOBs using JDBC and OGSA-DAI. OGSA-DAI client is making a SOAP call for each attribute value that is being inserted. These SOAP calls are the main portion of the overhead caused by OGSA-DAI.

id as an integer and the image as a BLOB object. The performance of JDBC and OGSA-DAI connections is shown in figure 3.

In case of JDBC, the images are converted into a byte array and inserted into the table. OGSA-DAI does not have any feature to insert binary data into the table directly. So, the data is pushed from the client into a stream, which in turn is sent to the data service resource. Here, OGSA-DAI is performing badly as the client is making a SOAP call to insert each attribute value (protein-protein interaction ID and image). The amount of time spent in these SOAP call is so high that the loading of images into local and remote databases is taking the equal amount of time. However, an interesting observation was made. As the size of the image is growing the difference between JDBC and OGSA-DAI performance is decreasing. If the images are of couple of MB, the difference between both of them is considerably decreased. The OGSA-DAI client is taking ca. 13 times more time to load the small number of images into a local database and 10 times more into a

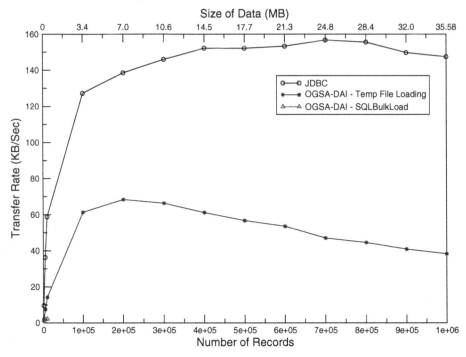

Fig. 4. Loading Oracle query results into a MySQL table of a remote database. The SQLBulkLoad is reporting connection timeout error after loading 4 MB data. So, the query results are written to a file and the file is used to upload into the table in remote database. See detailed explanation in text.

remote database than a normal JDBC client. However, when a large number of records were loaded the overhead is decreased to 8 and 6 times respectively.

3. **Loading query results of one database to a table in another database:** If the users would like to join a MySQL and an Oracle table the only way is to create a temp table of Oracle table in MySQL database and then join the temp table with MySQL table. OGSA-DAI is providing a feature called SQLBulkLoad to directly load the query results which are in web rowset format into another table. The Oracle query results were uploaded into a table in remote database. The performance of OGSA-DAI and JDBC is shown in figure 4.

The SQLBulkLoad was performing badly and after 4 MB size of data, it was reporting a connection time out error. To circumvent this problem the query results were written to a file which was loaded. The DeliverToFile would be an ideal activity for this but the DeliverToFile and DeliverFromFile activities are deactivated in our installation as they pose severe security threat. If they are activated, the user can access any file to which the globus user has access

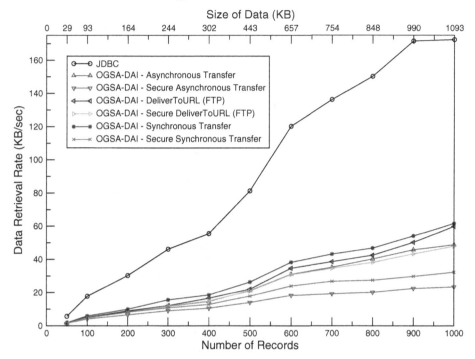

Fig. 5. Retrieving all information about the interaction partners including the sequences of a given SCOP domain, family, or superfamily id from a MySQL database. The database is on the same machine where globus and OGSA-DAI are deployed. Each record is of 736 Bytes size on average.

rights. So, DeliverToURL was used which can send the results using HTTP or HTTPs protocols or to a file on client using FTP protocol. The temporary file is transfered back using gsiscp to the OGSA-DAI server for loading. This is a full circle in moving the data but this seems to be only way how one can work around this problem in near future. The OGSA-DAI client requires almost 3 times more time than a normal JDBC client by writing the results to temp file and then uploading it into the table.

4.3 Retrieving Data

In this category, three scenarios were tested:

1. **Retrieving data from a MySQL database:** One of the important features of RNAi threading pipeline is to get the interaction partners given a SCOP domain or family or superfamily id. So, the query extracts all the information about the interactions like the sequences of interacting partners, details about interfaces, and information about the related literature.

Retrieving Data From A Remote Database

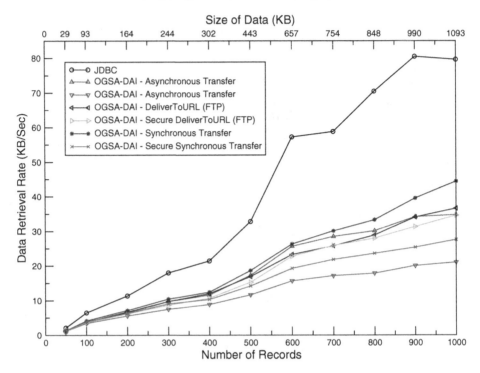

Fig. 6. Retrieving all the information about the interaction partners including the sequences of a given SCOP domain, family, or superfamily id from a MySQL database. The database is on a remote machine connected via WAN and deployed under OGSA-DAI. Each record is of 736 Bytes size on average.

The number of returned records depends on which SCOP family the user is querying. If he is querying an affluent family like Immunoglobulin over a hundred thousand records will be returned. But for most of the other SCOP families it will be a couple of hundred records. OGSA-DAI provides several delivery activities which can be used for such a case. In this scenario, data are retrieved with or without using transport and message level security, using different OGSA-DAI delivery options, and using non-secure JDBC connection. The performance of JDBC and OGSA-DAI on local and remote machines are shown in figure 5 and figure 6 respectively.

The important observation is the overhead introduced due to transport and message level security. In OGSA-DAI WSRF 2.1 version the encryption was one of the performance bottlenecks. However, the version 2.2 is far much better and the difference with and without security is negligible. The synchronous transfer is the fastest way to deliver the data to the client as all the results are fetched in one block. However, when a large amount of data is retrieved it will run out of memory. The tests revealed that ca. 3 MB data

Retrieving Large Amount of Data

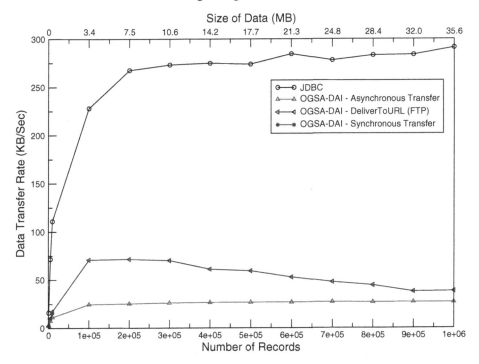

Fig. 7. Retrieving the atom coordinates and other PDB structure information from Macromolecular Structure Database (MSD), which is an oracle database. The database is on the same machine where globus and OGSA-DAI are deployed.

could be retrieved using synchronous transfer before running into memory problems. Thus, the DeliverToURL or asynchronous transfer are the only options to retrieve large amounts of data. In asynchronous transfer the query results are divided into blocks and each block is sent to the client asynchronously. To read large amount of data from the database the first choice should be to use DeliverToURL if FTP or GridFTP are available to use, otherwise one should use asynchronous data transfer. When the DeliverToURL activity was used, the resulting data were transformed into CSV format using XSL transformation before they were sent using FTP. To read from the local database the OGSA-DAI client (without or with transport and message level security) required in average about 3 to 4 times more time than a JDBC client, whereas to read from a remote database it requires 2 times more time than a JDBC client.

2. **Retrieving data from a Oracle database:** To check the performance of OGSA-DAI in reading large amount of data from the database, another Oracle query is considered. This query retrieves all three dimensional structure

Retrieving Images From Database

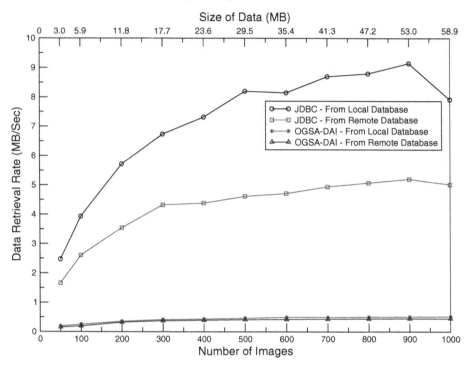

Fig. 8. Retrieving images from MySQL using JDBC connection and OGSA-DAI connection. Each image is on average ca. 61 KB. In case of OGSA-DAI, the images are transfered to the client using asynchronous mode as the synchronous mode could not transfer more than 50 images before throwing out of memory exception.

information given a PDB ID like atom coordinates from MSD database. The performance of OGSA-DAI and JDBC in this scenario are shown in figure 7. As mentioned earlier, the OGSA-DAI client is unable to retrieve more than 3 MB of data from the database as it will face memory problems. The OGSA-DAI client with DeliverToURL activity takes ca. 6 times more time than a normal JDBC client.

3. **Retrieving images from database:** If the RNAi pipeline user likes to see the interaction partners of a family with interface images, the only way to retrieve such information is through the asynchronous deliver activity. The synchronous deliver activity had thrown "out of memory" exception after reading about 50 images. The performance of OGSA-DAI and JDBC clients in this scenario is shown in figure 8. OGSA-DAI requires upto 12 to 18 times more time than a JDBC client to retrieve data from a remote or a local database respectively. The interesting observation is that the OGSA-DAI client is not able to retrieve the data or images with a rate more than 460 KB/sec either from local or remote database.

5   Conclusion

In this study OGSA-DAI WSRF 2.2 was evaluated and the performance was measured using scenarios that were adopted and thus comparable to a real application (RNAi screening).

Major advantages of OGSA-DAI for usage in a Grid infrastructure are:

- Access to databases occurs via the OGSA-DAI server. Thus, the databases (server firewalls) need to be opened only for the OGSA-DAI server(s). In a Grid environment, where an application can run on a large number of machines, this is the only way to provide better security for database providers.
- OGSA-DAI provides transport and message level security for data, which is a requirement for many medical applications and which other tools (e.g. JDBC) do not provide.

Since all the communication between OGSA-DAI server and the client is conducted through SOAP packages, accessing databases using OGSA-DAI is having some overhead when compared to a direct JDBC connection.

Nevertheless, the performance of OGSA-DAI in retrieving normal data types is quite satisfactory even with security and encryption. However, there is some space for improvement in asynchronous delivery activities. To achieve best OGSA-DAI performance, small amounts of data should be read using synchronous delivery and large amounts of data should be read using DeliverToURL with FTP protocol if its available otherwise using asynchronous delivery.

The loading and retrieving of images were included in the test scenario even though they are not that critical for the test application. But they are very important for other MediGRID applications, like medical image processing, which deal with patient information systems. Additionally, these applications require very granular authorization access rights. Using OGSA-DAI and storing the images in databases, one could ensure that only authorized persons can access the data at the attribute level granularity.

Acknowledgments

We are grateful to Thomas Steinke for providing the remote computing resources, Michael Schroeder, Christof Winter, and Andreas Henschel for providing the databases, Tim Mueller, and Patrick May for useful discussions. This work was funded by the German Federal Ministry of Education and Research (BMBF) within the D-Grid Integration and MediGRID projects.

References

[1] Antonioletti, M., Atkinson, M.P., Baxter, R., Borley, A., Hong, N.P.C., Collins, B., Hardman, N., Hume, A., Knox, A., Jackson, M., Krause, A., Laws, S., Magowan, J., Paton, N.W., Pearson, D., Sugden, T., Watson, P., Westhead, M.: The design and implementation of grid database services in ogsa-dai. Concurrency and Computation: Practice and Experience **17**(24) (February 2005) 357–376

[2] Winter, C., Henschel, A., Kim, W.K., Schroeder, M.: Scoppi: A structural classification of protein-protein interfaces. Nucleic Acids Research (Accepted 2005)

[3] May, P., Steinke, T.: Theseus - protein structure prediction at zib. In preparation (2006)

[4] May, P., Steinke, T.: THESEUS - protein structure prediction at ZIB. ZIB Report **06-24** (2006)

[5] May, P., Ehrlich, H.C., Steinke, T.: ZIB Structure Prediction Pipeline: Composing a Complex Biological Workflow through Web Services. In Nagel, W.E., Walter, W.V., Lehner, W., eds.: Euro-Par 2006 Parallel Processing. LNCS 4128, Springer Verlag (2006) 1148-1158

[6] Murzin, A.G., Brenner, S.E., Hubbard, T., Chothia, C.: Scop: A structural classification of proteins database for the investigation of sequences and structures. Journal of Molecular Biology **247**(4) (April 1995) 536-540

[7] Doms, A., Schroeder, M.: Gopubmed: Exploring pubmed with the geneontology. Nucleic Acids Research **33** (July 2005) 783-786

[8] Huang, B., Schroeder, M.: Using residue propensities and tightness of fit to improve rigid-body protein-protein docking. Proceedings of German Bioinformatics Conference **GI LNI71** (2005)

[9] Rajasekar, A., Wan, M., Moore, R., Schroeder, W., Kremenek, G., Jagatheesan, A., Cowart, C., Chen, S.Y., Olaschanowsky, R.: Storage resource broker - managing distributed data in a grid. J. Comput. Soc, India **33**(4) (December 2003) 41-53

Systems Support for Remote Visualization of Genomics Applications over Wide Area Networks

Lars Ailo Bongo[1], Grant Wallace[2], Tore Larsen[1],
Kai Li[2], and Olga Troyanskaya[2,3]

[1] Department of Computer Science, University of Tromsø, N-9037 Tromsø, Norway
[2] Department of Computer Science, Princeton University, Princeton NJ 08544, USA
[3] Lewis-Sigler Institute for Integrative Genomics, Princeton University,
Princeton NJ 08544, USA
{larsab, tore}@cs.uit.no
{gwallace, li, ogt}@cs.princeton.edu

Abstract. Microarray experiments can provide molecular-level insight into a variety of biological processes, from yeast cell cycle to tumorogenesis. However, analysis of both genomic and protein microarray data requires interactive collaborative investigation by biology and bioinformatics researchers. To assist collaborative analysis, remote collaboration tools for integrative analysis and visualization of microarray data are necessary. Such tools should: (i) provide fast response times when used with visualization-intensive genomics applications over a low-bandwidth wide area network, (ii) eliminate transfer of large and often sensitive datasets, (iii) work with any analysis software, and (iv) be platform-independent. Existing visualization systems do not satisfy all requirements. We have developed a remote visualization system called Varg that extends the platform-independent remote desktop system VNC with a novel global compression method. Our evaluations show that the Varg system can support interactive visualization-intensive genomic applications in a remote environment by reducing bandwidth requirements from 30:1 to 289:1.

Keywords: Remote visualization, genomics collaboration, Rabin fingerprints, compression.

1 Introduction

Interactive analysis by biology and bioinformatics researchers is critical in extracting biological information from both genomic [1], [2] and proteomic [3], [4], [5], [6], [7] microarrays. Many supervised and unsupervised microarray analysis techniques have been developed [8], [9], [10], [11], and the majority of these techniques share a common need for visual, interactive evaluation of results to examine important patterns, explore interesting genes, or consider key predictions and their biological context.

Such data analysis in genomics is a collaborative process. Most genomics studies include multiple researchers, often from different institutions, regions, and countries. Of the 20 most relevant papers returned by BioMed Central with the query

W. Dubitzky et al. (Eds.): GCCB 2006, LNBI 4360, pp. 157–174, 2007.
© Springer-Verlag Berlin Heidelberg 2007

"microarray," 14 had authors located at more than one institution, and 7 had authors located on either different continents or cross continents. Such collaboration requires interactive discussion of the data and its analysis, which is difficult to do without sharing a visualization of the results. To make such discussions truly effective, one in fact needs not just static images of expression patterns, but an opportunity to explore the data interactively with collaborators in a seamless manner, independent of the choice of data analysis software, platforms, and of researchers' geographical locations.

We believe that an ideal collaborative, remote visualization system for genomic research should satisfy three requirements. First, synchronized remote visualization should have a fast response time to allow collaborating parties to interact smoothly, even when using visualization-intensive software across a relatively low-bandwidth wide area network (WAN). Second, collaborating parties should not be required to replicate data since microarray datasets can be large, sensitive, proprietary, and potentially protected by patient privacy laws. Third, the system should allow collaborators to use any visualization and data analysis software running on any platform.

Existing visualization systems do not satisfy all three requirements above. Applications with remote visualization capabilities may satisfy the first and the second requirements, but typically not the third as require universal adoption among participating collaborators. Thin-client remote visualization systems, such as VNC [12], Sun Ray [13], THINC [14], Microsoft Remote Desktop [15] and Apple Remote Desktop [16] satisfy only the second requirement because they do not perform intelligent data compression and all except VNC are platform-dependent. Web browser-based remote visualization software can satisfy the third requirement, but not the first two because these systems are not interactive and do not optimize the network bandwidth requirement.

This paper describes the design and implementation of a remote visualization system called Varg that satisfies all three requirements proposed above. To satisfy the first requirement, the Varg system implements a novel method to compress redundant two-dimensional pixel segments over a long visualization session. To satisfy the second and the third requirements, the Varg system is based on a platform-independent remote desktop system VNC, whose implementation allows remote visualization of multiple applications in a network environment.

The main contribution of the Varg system is the novel method for compressing 2-D pixel segments for remote genomic data visualization. Genomic data visualization has two important properties that create opportunities for compression. The first is that datasets tends to be very large. A microarray dataset typically consists of a matrix of expression values for thousands or tens of thousands of genes (rows). The second is that due to the limitation of display scale and resolution, researchers typically view only tens of genes at a time by frequently scrolling visualization frames up and down. As a result, the same set of pixels will be moved across the display screen many times during a data analysis and visualization session. We propose a novel method to identify, compress and cache 2-D pixels segments. Not sending redundant segments across the WAN greatly improves the effective compression ratio reducing network bandwidth requirements for remote visualization.

Our initial evaluation shows that the prototype Varg system can compress display information of multiple genomic visualization applications effectively, typically reducing the network bandwidth requirement by two orders of magnitude. We also demonstrate that this novel method is highly efficient and introduces a minimal overhead to the networking protocol; and that the Varg system can indeed support multiple visualization-intensive genomic applications in a remote environment interactively with minimal network bandwidth requirement.

2 System Overview

Varg is a network bandwidth optimized, platform-independent system that allows users to interactively visualize multiple remote genomic applications across a WAN. The architecture of Varg is based on a client-server model as shown in Fig. 1. Varg leverages the basic VNC protocol (called RFB) to implement platform-independent remote visualization and extends it with a high-speed 2-D pixel segment compression module with a cache in the server and a decompression module with a cache in the client. The Varg server runs multiple visualization applications, compresses their two-dimensional pixel segments, and communicates with the remote Varg client. The client decompresses the data utilizing a large cache and performs remote visualization.

The caches of the Varg server and client cooperate to minimize the required network bandwidth by avoiding redundant data transfers over the network. Unlike other global compression methods for byte data streams [17, 18], Varg is designed to optimize network bandwidth for remote data transfers of 2-D pixels segments generated by genomic visualization applications on the VNC server.

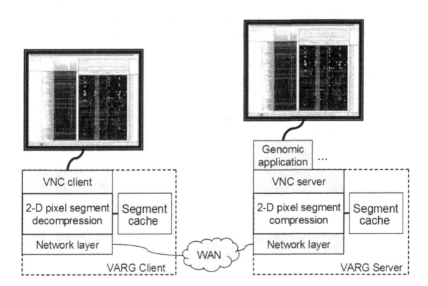

Fig. 1. Remote visualization overview. The VNC server sends screen updates to the VNC client. The Varg system caches updates and provides compression by replacing updates already in the client cache with the cache index.

Since Varg is built on the VNC protocol, it allows multiple users to conveniently visualize and control a number of applications in a desktop across a network. When an owner of some sensitive or very large data set wants to collaborate with a remote collaborator, she can run one or more analysis programs that access her sensitive data on her Varg server, which connects with a Varg client on her collaborator's site. The researchers can then use these programs in a synchronized fashion across the network. Although the collaborator can visualize and control the application programs in the same way as the owner, the Varg client receives only visualization pixels from the Varg server; no sensitive data is ever transferred across the network. We expect that this feature may be especially useful to researchers working with clinical data due to privacy and confidentiality concerns.

3 Compressing 2-D Pixel Segments

The main idea in the Varg system is to compress visualization pixel data at a fine-grained 2-D pixel segment level. The system compresses 2-D pixel segments by using a global compression algorithm to avoid sending previously transferred segments and by applying a slow, but efficient, local compression [19] on the unique segments. This section describes Varg's basic compression algorithm, explains our novel content-based anchoring algorithm for 2-D pixel segments, and outlines an optimization of the compression algorithm using a two-level fingerprinting scheme that we developed.

3.1 Basic Compression Algorithm

The basic compression algorithm uses fingerprints together with cooperative caches on the Varg server and client to identify previously transferred pixel segments, as shown in Fig. 2.

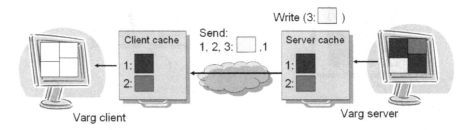

Fig. 2. Compression scheme. The screen is divided into regions which are cached at both ends of the low-bandwidth network. Fingerprints are sent in place of previously sent regions.

The algorithm on the Varg server is:

- Process an updated region of pixels from the VNC server
- Segment the region into 2-D pixel segments

- For each segment, compute its fingerprint and use the fingerprint as the segment's identifier to lookup in the server cache. If the segment has not been sent to the Varg client previously, compress the segment with a local compression method and send the segment to the client. Otherwise, send the fingerprint instead.

The algorithm on the Varg client is:

- If the received data is a 2-D pixel segment, decompress it with a corresponding algorithm, write the fingerprint and segment to the cache, and then pass the segment to the VNC client
- If the received data is a fingerprint, retrieve the segment of the fingerprint from its cache and then pass the segment to the VNC client.

The basic algorithm is straightforward and its high-level idea is similar to previous studies on using fingerprints (or secure hashes) as identifiers to avoid transfer of redundant data segments [20, 21], [22], [17], [18]. The key difference is that previous studies are limited to deal with one-dimensional byte streams and have not addressed the issue of how to anchor 2-D pixel segments. In a later section, we will also present an algorithm to use short fingerprints to compress repeated 2-D pixel segments.

3.2   Content-Based Anchoring of 2-D Pixel Segments

One basis of our approach is content-based anchoring where the 2-D region of display-pixels is divided into segments based on segment content. A simpler approach would be to anchor segments statically (such as an 8×8 pixel grid, used in MPEG compression algorithms). The problem with a static approach is that the anchoring is sensitive to screen scrolls. When a user scrolls her visualization by one pixel, the segmentation of 2-D pixels will be shifted by one pixel relative to the displayed image. Even if the entire scrolled screen has been transferred previously, the content of segments will typically have changed, giving a new fingerprint and requiring a new transfer across the WAN.

Our approach is to perform content-based anchoring instead of static anchoring. The anchoring algorithm takes its input from the frame-buffer, and returns a set of rectangular segments which subdivide the screen. The goal of the algorithm is to consistently anchor the same groups of pixels no matter where they are located on the screen. The main difficulty in designing a content-based anchoring algorithm for a screen of pixels is that the data is two dimensional.

Manber introduced a content-based technique to anchor one-dimensional data segments for finding similar files [22]. His method applies a Rabin fingerprint filter [23] over a byte data stream and identifies anchor points wherever the k least significant bits of the filter output are zeros. With a uniform distribution, an anchor point should be selected every 2^k bytes.

Our algorithm combines the statically divided screen approach with Manber's technique. The algorithm is based on the observation that content motion in microarray analysis is often due to vertical or horizontal scrolling. However, it is not practical to do redundancy detection both horizontally and vertically due to the computational cost and reduced compression ratio caused by overlapping regions.

Therefore, we estimate whether the screen has moved mostly horizontally or mostly vertically using Manber's technique. We generate representative fingerprints for every 32nd row, and every 32nd column for the screenshot, and compare how many fingerprints are similar to the row and column fingerprints of the previous screenshot. Assuming that horizontal scrolling or moving will change most row fingerprints, but only a few column fingerprints, we can compare the percentage of similar row and column fingerprints to estimate which movement is dominant.

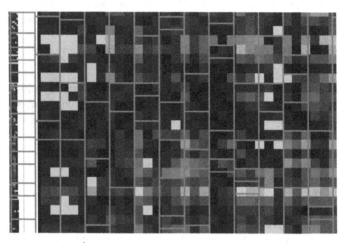

Fig. 3. A portion of the screen that is divided into segments that move with the content

For predominately vertical motion we statically divide the screen into m columns (m times screen height) and divide each column into regions by selecting anchoring rows. The anchoring rows are selected based on their fingerprint calculated using a four byte at a time Rabin fingerprint implementation. The column segmentation is ideal for scrolling because the regions move vertically with the content. If we detect predominately horizontal motion instead, we run the same algorithm but divide the screen into rows first and then divide each row into regions by selecting anchoring columns.

Screen data can include pathological cases when large regions of the screen have the same color. For such regions, the fingerprints will be identical. Thus, either all or no fingerprints will be selected. To avoid such cases, our algorithm does fingerprint selection in three steps. First all fingerprints are calculated. Second, we scan the fingerprints and mark fingerprints as *similar* if at least s subsequent fingerprints are identical. Third, we select fingerprints using the k most significant bits, while imposing a minimum distance m between selected fingerprints. Also, the first and last rows are always selected.

Empirically we have found that the best results are achieved for $s = 8$, $m = 16$ or 32, and k such that each 64th row on the average is selected. Also, such similar regions compress well using a local compression algorithm such as zlib [24] due to their repeated content. We have found empirically that imposing a maximum distance does not improve the compression ratio or compression time.

3.3 An Optimization with Two-Level Fingerprinting

An important design issue in using fingerprints as identifiers to detect previously transferred data segments is the size of a fingerprint. Previous systems typically chose a secure hash, such as 160-bit SHA-1 [25], as a fingerprint so that the probability of a fingerprint collision can be lower than a hardware bit error rate. However, since the global compression ratio is limited to the ratio of the average pixel segment size to the fingerprint size, increasing the fingerprint size reduces this limit on the compression ratio.

To maximize the global compression ratio and maintain a low probability of fingerprint collision, we use a two-level fingerprinting strategy. The low-level fingerprinting uses 32-bit Rabin fingerprint of fingerprints, one for each 2-D pixel segments. Although using such short fingerprints will have a higher probability of a fingerprint collision, they can be computed quickly using the fingerprints already computed for the anchoring, thereby maintaining a high global compression ratio.

The high-level fingerprinting uses SHA-1 hashes as fingerprints. It computes a 160-bit fingerprint for each of the transferred pixel segments. The server computes such a long fingerprint as a strong checksum to detect low-level fingerprint collisions. When a low-level fingerprint collision is detected, the server resends the pixel segment covered by the long fingerprint.

Another way to look at this method is that the server may send two sets of updates, the first based on short fingerprints that can have collisions, and the second set of updates consisting of corrections in case of short fingerprint collisions. This method reduces the user perceived end-to-end latency.

4 Implementation

We have implemented a prototype system (called Varg) consisting of a sequential server and a client, as described in Section 2. The Varg server implements the 2-D pixel segment compression algorithm and Varg client implements the corresponding decompression algorithm described in the previous section.

The integration of Varg compression, decompression, and cache modules with the VNC client and server are simple. VNC has only one graphics primitive: "Put rectangle of pixels at position (x, y)" [12]. This allows separating the processing of the application display commands from the generation of display updates to be sent to the client. Consequently the server only needs to detect updates in the frame-buffer, and can keep the client completely stateless.

Varg employs a synchronized client and server cache architecture that implements an eventual consistency model using the two-level fingerprinting mechanism. The client and server caches are initialized at Varg system start time. The client cache is then synchronized by the updates sent from the server. The compression algorithm requires the client cache to maintain the invariant that whenever the client receives a fingerprint, its cache must have the fingerprint's segment. Since short fingerprints may have collisions, our prototype allows the client cache to contain any segment of the same short fingerprint at a given time. The long fingerprint will eventually trigger an update to replace it with a recently visualized segment.

5 Evaluation

We have conducted an initial evaluation of the Varg prototype system. The goal of the evaluation is to answer the following two questions:

- What are the network communication requirements for remote visualization of genomic applications?
- How much compression of network communication data can the Varg prototype system achieve for remote visualization of genomic applications?

To answer the first question, we have measured the difference between available bandwidth on current WANs and the required bandwidth for remote visualization of Genomic applications. To answer the second question, we used a trace-driven VNC emulator to find how much the Varg system can reduce the communication time for three genomic applications. In the rest of this section, we will present our experimental testbed and then our evaluation results to answer each question.

5.1 Experimental Testbed

In order to compare compression ratios of various compression algorithms, our experimental testbed (Fig. 4) employs two identical Dell Dimension 9150, each with one dual-core 2.8 GHz Pentium D processor and 2 GB of main memory. Both computers run Fedora Core 4, with Linux kernel 2.6.17SMP.

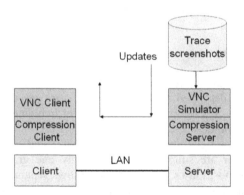

Fig. 4. Experimental testbed for the bandwidth requirements and compression ratio evaluation

The server runs with a screen resolution of 1280x1024 pixels and with a color depth of 32 bits per pixel. We also run an experiment on a display wall with a resolution of 3328x1536 pixels.

To compare different systems, an important requirement is to drive each system with the same remote visualization workloads. To accomplish this goal, we have used a trace-driven approach. To collect realistic traces, we used the Java AWTEventListener interface to instrument three genomic microarray analysis applications. We used these to record a 10-minute trace containing all user input events for each case. Later the traces were used to create a set of screenshots, each taken after playing back a recorded mouse or keyboard event that changes the screen content. The screenshots are used by a VNC simulator that copies a screenshot to a shadow framebuffer, and invokes the Varg server, which does change detection and compression before sending the updates to the client.

5.2 Network Communication Requirements

In order to answer the question about the network communication requirements for remote visualization of genomic applications, we need to answer several related questions including the composition of communication overhead, the characteristics of available networks, the behavior of remote visualization of genomic applications, and the required compression ratio to meet certain interactive requirements. Our finding is that genomic applications require high compression ratio to compress the pixel data to use existing WAN connections.

The network communication overhead can be expressed with a simple formula:

$$2L + \frac{S}{B \cdot R} + C \tag{1}$$

where L is the network latency, S is the data to be transferred, B is the network bandwidth, R is compression ratio, and C is compression time. The formula considers compression a part of the network communication mechanism, thus the total communication overhead includes the round-trip network latency plus the time to compress and transfer the data. This formula ignores the overheads of several software components such as the VNC client and server. Also, we usually ignore decompression time since it is low compared to the compression time (less than 1msec).

Based on this formula, it is easy to see that different network environments have different implications for remote visualization. Conventional wisdom assumes that WANs have low bandwidth. To validate this assumption we used Iperf [26] to measure the TCP/IP throughput between a server and a client connected using various local and wide area networks. The following table shows that the WAN throughput ranges from 0.2 to 2.13 Mbytes/sec (Table 1). This is up to 400 times lower than for Gigabit Ethernet. Also, the two-way latency is high, ranging from 11—120 ms.

Table 1. TCP/IP bandwidth and latency for client-server applications run on local area and wide area networks

Network	Bandwidth (Mbytes/sec)	Latency (msec)
Gigabit Ethernet	80.00	0.2
100 Mbps Ethernet	8.00	0.2
Princeton – Boston	2.13	11
Princeton – San Diego	0.38	72
Princeton (USA(– Tromsø (Norway)	0.20	120

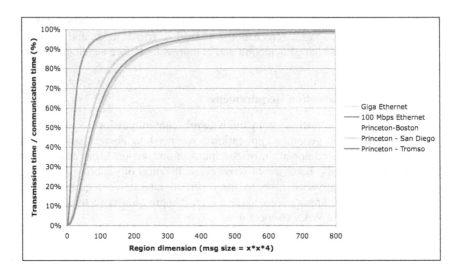

Fig. 5. For regions larger than 80×80 pixels, the transmission time dominates the total communication overhead

Based on the characteristics of the available networks, an interesting question is what size of network transfers contribute significantly to the total communication overhead. Fig. 5 shows how much transmission contributes to the communication time depending on the amount of pixel data sent over the network connection. For all WAN networks, the ratio of transmission time to communication time is more than 50% for regions more than about 80×80 pixels or 25 Kbytes.

Two natural questions are, how frequent are screen updates larger than 80×80 pixels for genomics applications, and are the update sizes different compared to Office applications usually used in remote collaboration. To answer these questions, we measured the average VNC update size for three sessions using three applications on Windows XP:

1. Writing this paper in Microsoft Word.
2. Preparing the figures for this paper in Microsoft PowerPoint.
3. Microarray analysis using the popular Java Treeview software [27].

For each application, we recorded a session lasting about 10 minutes. We instrumented the VNC client to record the time and size of all screen updates received. We correlated these to when the update requests were sent, to get an estimate for the size of each screen update.

The results show that updated regions are much larger for the genomic application than for the two office applications (Fig. 6). About 50% of the messages are larger than 80×80 pixels, and hence for these the transmission time will be longer than the network latency for the WANs. Another observation is that the genomic application has a higher update frequency than office applications. Combined these increase the required bandwidth.

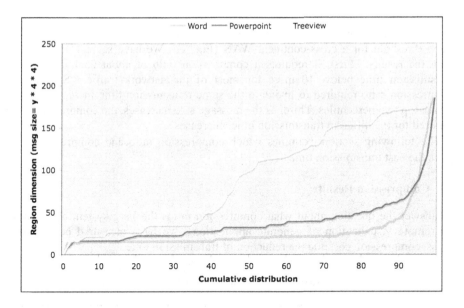

Fig. 6. The screen regions update sizes for the Java Treeview application are much larger than for the Office applications. About 50% of the messages are more than 80x80 pixels.

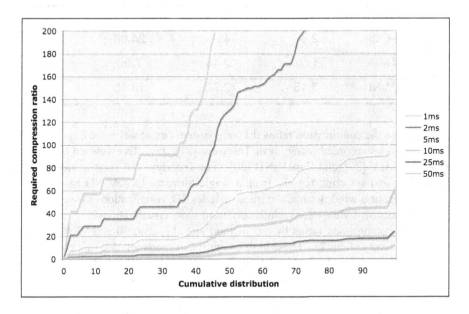

Fig. 7. Compression ratio required to keep transmission overhead below a given threshold for the Princeton-San Diego network connection. The x-axis shows the percentiles for the Treeview message sizes in Fig. 6. Compression time is not taken into account.

To see the impact of compressing pixel data for remote visualization, we have calculated the compression ratio necessary to maintain the transmission time below a given threshold for a cross-continent WAN (Fig. 7). We have several observations from the results. First, it requires a compression ratio of about 25:1 to keep the transmission time below 10 msec for most of the network traffic. Second, the compression ratio required to maintain the same transmission time increases rapidly for the top two percentiles. Third, as the message size increases, the compression ratio required for the different transmission times increases.

The following section examines which compression ratio and compression time gives the best transmission time.

5.3   Compression Results

To answer the question about what compression ratios the Varg system can achieve for remote visualization of genomic applications, we have measured compression ratios, compression cost and the reduction of transmission time.

Table 2. Compression ratio for four genomic data analysis applications

	Differencing	2D pixel segment compression	Ziv-Lempel (zlib)	Total compression
TreeView	1.89	5.74	19.98	216.76
TreeView-Cube	2.87	4.05	24.88	289.19
TMeV	1.52	2.46	7.90	29.54
GeneVaND	3.15	2.72	10.85	92.96

To measure the compression ratios the Varg system can achieve, we have used four 15-minute traces recorded using: Java Treeview [27], Java Treeview on the display wall [28], TMeV [29], and GeneVaND [30]. For Treeview, the visualizations mostly are scrolling and selecting regions from a single bitmap. GeneVaND has relatively small visualization windows and the trace includes 3D visualizations as well as some 2D visualizations. TMeV trace includes different short visualizations.

The total compression ratios by our method are 217, 289, 30 and 93 for the four traces respectively (Table 2). These high compression ratios are due to a combination of three compression methods: Region differencing, 2D pixel segment compression, and zlib local compression. We have several observations based on these data. First, the combined compression results are excellent. Second, zlib contributes the most in all cases, but zlib alone is not enough to achieve high compression ratios. Third, the 2D pixel segment compression using fingerprinting contributes fairly significantly to the compression ratio ranging from 2.5 to 5.7. This is due to the fact that the differencing phase has already removed a large amount of redundant segments.

Table 3. Average compression time per screen update. The total compression time depends on the application window size, and how well the differencing and 2D pixel segment compression modules compress the data before zlib is run.

	Differencing	2D pixel segment compression	Ziv-Lempel (zlib)	SHA-1
TreeView	0.9 ms	3.8 ms	11.1 ms	3.5 ms
TreeView-Cube	2 ms	7.9 ms	30.2 ms	7 ms
TMeV	1.3 ms	6.6 ms	83.4 ms	7.7 ms
GeneVaND	1 ms	2.7 ms	10.1 ms	1.5 ms

To understand the contribution of different compression phases to the compression time, we measured the time spent in each module (Table 2). The most significant contributor is zlib, which consumes more than 10 ms in all cases. In TMeV it consumes more than 83ms, since more data is sent through this module due to the low compression ratios for the differencing and 2D pixel segment compression modules. The second most significant contributor is anchoring, but it is below 8 ms even for the display wall case. Although SHA-1 calculation contributes up to 8ms in the worst case, its computation overlaps with network communication.

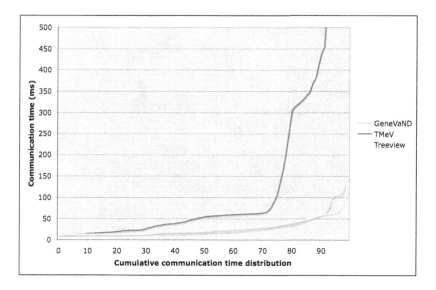

Fig. 8. Communication time distribution for update messages over the Princeton—Boston network. For Treeview and GeneVaND more than 90% of the communication overheads are less than 100ms. The update size distribution differs from Fig. 5, since a more accurate tracing tool was used to capture the trace.

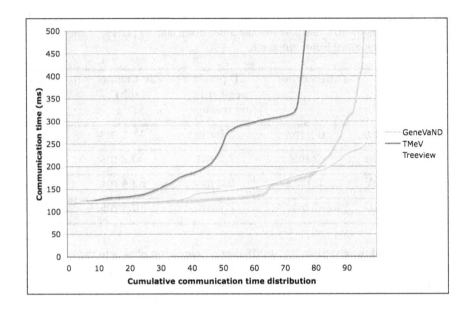

Fig. 9. Communication time distribution for the Princeton—Tromsø network. For Treeview and GeneVaND more than 80% of the communication overheads are less than 200ms.

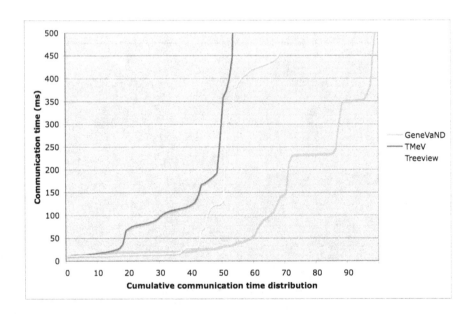

Fig. 10. Communication time distribution with VNC compression for the Princeton—Boston network. Compared to Varg the communication time shown in Fig. 9 significantly increases for large messages.

To understand the reduction in communication time, we recorded for each update the number of compressed bytes returned by each module, and the compression time for each module. This allows us to use Formula 1 to estimate the communication time for each of the WANs in Table 1. The cumulative distribution of communication times for the highest and lowest bandwidth networks are shown in Fig. 8 and Fig. 9. Without compression the communication overhead for the largest updates is several seconds. For the Princeton—Boston network the communication overhead with Varg is less than 100ms for over 90% of the messages (except for TMeV). On the Princeton—Tromsø network, for 80% of the update operations the communication overhead is less than 200ms, of which the latency contributes to 112 ms.

To compare our compression against the zlib compression used in many VNC implementations for low-bandwidth networks, we disabled the 2D pixel segment compression module in Varg, and did a similar calculation as above (Fig. 10). The results show a significant increase in communication time, especially for Treeview where the communication overhead is more than 300ms for about 50% of the messages.

6 Related Work

Compression algorithms used by VNC [12] implementations either take advantage of neighboring region color similarities, use general purpose image compression [31] such as JPEG [32], or general purpose compression such as zlib [19]. Neighboring region redundancy compression is fast but has low compression ratio. Therefore zlib is usually used for WANs. Our results show that the compression time for zlib is high. JPEG is lossy, and is not suited for Microarray analysis, since it may introduce visual artifacts that may influence the biologist's interpretation of the data.

Remote visualization systems that use high level graphics primitives for communication, such as Microsoft Remote Desktop [15], are able to cache bitmaps used for buttons and other GUI components. However, the high-level graphics primitives do not compress well leading to performance problems in WANs [13, 33].

Encoders used for streaming video, such as MPEG [34], compress data by combining redundancy detection and JPEG type compression. Usually a static pixel grid is used, which we have shown gives worse performance than our approach. In addition the MPEG compression is lossy and there are no real time encoders available. TCC-M [35] is a block movement algorithm designed for thin-client visualization that use unique pixels in an image (feature sets) to detect 2D region movement. However, redundancy is only detected between the two latest screen updates thus reducing the compression ratio.

Earlier one-dimensional fingerprinting approaches [17, 18] require the two-dimensional screen to be converted to some one-dimensional representation. This will split up two-dimensional regions on the screen causing the size of the redundant regions to decrease, hence reducing the compression ratio.

Access Grid [36] provides multiple collaborators with multimedia services by multicasting audio, video and remote desktop displays such as VNC. However, Access Grid does not provide compression to reduce the network bandwidth requirement for specific data visualization such as genomic data exploration. Since

Varg extends the VNC protocols to compress 2D segments for genomic data visualization, it can effectively work together with Access Grid systems to support multi-party collaborations.

7 Conclusion

This paper presents the design and implementation of the Varg system: a network bandwidth optimized, platform-independent system that allows users to interactively visualize multiple remote genomic applications across a WAN. The paper has proposed a novel method to compress 2-D pixel segments by using fingerprinting and proposed a two-level fingerprinting method to improve global compression, and to reduce compression time.

We also found that genomic applications have much higher network bandwidth requirements than office applications. They require substantial compression of network data to achieve interactive remote data visualization on some examples of existing WAN.

An initial evaluation of our prototype system shows that the proposed 2-D pixel segment compression method works well and imposes only modest overheads. By combining with zlib and differencing compression methods, the prototype system achieved compression ratios ranging from 30:1 to 289:1 for four genomic visualization applications that we have experimented with. Such compression ratios allow the Varg system to run remote visualization of genomic data analysis applications interactively across WANs with relatively low available network bandwidths.

Acknowledgments

This work was done while LAB and TL were visiting Princeton University and was supported in parts by Princeton University, The University of Tromsø, The Research Council of Norway (incl. project No. 164825), NSF grants CNS-0406415, EIA-0101247, CNS-0509447, CCR-0205594, and CCR-0237113, and NIH grant R01 GM071966. OGT is an Alfred P. Sloan Research Fellow

References

1. Lipshutz RJ, Fodor SPA, Gingeras TR, Lockhart DJ: **High density synthetic oligonucleotide arrays**. *Nature Genetics* 1999, **21**:20-24.
2. Schena M, Shalon D, Davis RW, Brown PO: **Quantitative Monitoring of Gene-Expression Patterns with a Complementary-DNA Microarray**. *Science* 1995, **270**(5235):467-470.
3. Cahill DJ, Nordhoff E: **Protein arrays and their role in proteomics**. *Adv Biochem Eng Biotechnol* 2003, **83**:177-187.
4. Sydor JR, Nock S: **Protein expression profiling arrays: tools for the multiplexed high-throughput analysis of proteins**. *Proteome Sci* 2003, 1(1):3.

5. Oleinikov AV, Gray MD, Zhao J, Montgomery DD, Ghindilis AL, Dill K: **Self-assembling protein arrays using electronic semiconductor microchips and in vitro translation.** *J Proteome Res* 2003, **2**(3):313-319.

6. Huang RP: **Protein arrays, an excellent tool in biomedical research.** *Front Biosci* 2003, **8**:d559-576.

7. Cutler P: **Protein arrays: the current state-of-the-art.** *Proteomics* 2003, **3**(1):3-18.

8. Kerr MK, Churchill GA: **Bootstrapping cluster analysis: assessing the reliability of conclusions from microarray experiments.** *Proc Natl Acad Sci U S A* 2001, **98**(16):8961-8965.

9. Yeung KY, Haynor DR, Ruzzo WL: **Validating clustering for gene expression data.** *Bioinformatics* 2001, **17**(4):309-318.

10. Mendez MA, Hodar C, Vulpe C, Gonzalez M, Cambiazo V: **Discriminant analysis to evaluate clustering of gene expression data.** *FEBS Lett* 2002, **522**(1-3):24-28.

11. Datta S, Datta S: **Comparisons and validation of statistical clustering techniques for microarray gene expression data.** *Bioinformatics* 2003, **19**(4):459-466.

12. Richardson T, Stafford-Fraser Q, Wood KR, Hopper A: **Virtual network computing.** *Ieee Internet Computing* 1998, **2**(1):33-38.

13. Schmidt BK, Lam MS, Northcutt JD: **The interactive performance of SLIM: a stateless, thin-client architecture.** In: *Proceedings of the seventeenth ACM symposium on Operating systems principles.* Charleston, South Carolina, United States: ACM Press; 1999.

14. Baratto RA, Kim LN, Nieh J: **THINC: a virtual display architecture for thin-client computing.** In: *Proceedings of the twentieth ACM symposium on Operating systems principles.* Brighton, United Kingdom: ACM Press; 2005.

15. Cumberland BC, Carius G, Muir A: **Microsoft Windows NT Server 4.0, Terminal Server Edition: Technical Reference.** In. Edited by Press M. Redmond, WA; 1999.

16. **Apple Remote Desktop** [http://www.apple.com/remotedesktop/]

17. Spring NT, Wetherall D: **A protocol-independent technique for eliminating redundant network traffic.** In: *Proceedings of the conference on Applications, Technologies, Architectures, and Protocols for Computer Communication.* Stockholm, Sweden: ACM Press; 2000.

18. Muthitacharoen A, Chen B, Mazières D: **A low-bandwidth network file system.** In: *Proceedings of the eighteenth ACM symposium on Operating systems principles.* Banff, Alberta, Canada: ACM Press; 2001.

19. Ziv J, Lempel A: **A universal algorithm for sequential data compression.** *IEEE Transactions on Information Theory* 1977 **23**(3):337 - 343.

20. Broder A: **Some applications of Rabin's fingerprinting method.** In: *Sequences II: Methods in Communications, Security, and Computer Science: 1993*; 1993.

21. Broder A: **On the Resemblance and Containment of Documents.** In: *Proceedings of the Compression and Complexity of Sequences 1997.* IEEE Computer Society; 1997.

22. Manber U: **Finding similar files in a large file system.** In: *Proceedings of the Winter 1994 USENIX Technical Conference.* San Francisco, CA; 1994.

23. Rabin MO: **Fingerprinting by random polynomials.** In: *Technical Report TR-15-81.* Center for Research in Computing Technology, Harvard University; 1981.

24. **DEFLATE Compressed Data Format Specification version 1.3.** In: *RFC 1951.* The Internet Engineering Task Force; 1996.

25. **Secure Hash Standard.** In: *FIPS PUB 180-1.* National Institute of Standards and Technology; 1995.

26. **Iperf webpage** [http://dast.nlanr.net/Projects/Iperf/]

27. Saldanha AJ: **Java Treeview--extensible visualization of microarray data.** *Bioinformatics* 2004, **20**(17):3246-3248.
28. Wallace G, Anshus OJ, Bi P, Chen HQ, Clark D, Cook P, Finkelstein A, Funkhouser T, Gupta A, Hibbs M *et al*: **Tools and applications for large-scale display walls.** *Ieee Computer Graphics and Applications* 2005, **25**(4):24-33.
29. Saeed AI, Sharov V, White J, Li J, Liang W, Bhagabati N, Braisted J, Klapa M, Currier T, Thiagarajan M *et al*: **TM4: a free, open-source system for microarray data management and analysis.** *Biotechniques* 2003, **34**(2):374-378.
30. Hibbs MA, Dirksen NC, Li K, Troyanskaya OG: **Visualization methods for statistical analysis of microarray clusters.** *BMC Bioinformatics* 2005, **6**:115.
31. Richardson T: **The RFB Protocol version 3.8.** In.: RealVNC Ltd; 2005.
32. Gregory KW: **The JPEG still picture compression standard.** *Commun ACM* 1991, **34**(4):30-44.
33. Lai AM, Nieh J: **On the performance of wide-area thin-client computing.** *ACM Trans Comput Syst* 2006, **24**(2):175-209.
34. Gall DL: **MPEG: a video compression standard for multimedia applications.** *Commun ACM* 1991, **34**(4):46-58.
35. Christiansen BO, Schauser KE: **Fast Motion Detection for Thin Client Compression.** In: *Proceedings of the Data Compression Conference (DCC '02).* IEEE Computer Society; 2002.
36. **Access Grid** [http://www.accessgrid.org/]

HVEM DataGrid: Implementation of a Biologic Data Management System for Experiments with High Voltage Electron Microscope

Im Young Jung[1], In Soon Cho[1], Heon Y. Yeom[1],
Hee S. Kweon[2], and Jysoo Lee[3]

[1] School of Computer Science and Engineering,
Seoul National University, Seoul 151-742, Korea
{iyjung,ischo,yeom}@dcslab.snu.ac.kr
[2] Electron Microscopy Team, KBSI, Taejon, Korea
hskweon@kbsi.re.kr
[3] Supercomputing Center, KISTI, Taejon, Korea
jysoo@kisti.re.kr

Abstract. This paper proposes High Voltage Electron Microscope (HVEM) DataGrid for biological data management. HVEM DataGrid allows researchers to share the results of their biological experiments using HVEM, so that they can analyze them together to perform good research. The proposed system is for people whose primary work is to access HVEM, to obtain experimental results for biological samples and to store or retrieve them on HVEM DataGrid. The architecture of the HVEM Grid[3] is designed to materialize all the necessary conditions in allowing the user 1) to control every part of HVEM in a fine-grained manner, 2) to check the HVEM and observe various states of the specimen, 3) to manipulate their high resolution 2-D and 3-D images, and 4) to handle the experimental data including storing and searching them. HVEM DataGrid, a subsystem of the HVEM Grid system, is to provide a simple web-based controlling method for remote biological data.

1 Introduction

Many areas of science currently use computing resources as part of their research and try to utilize more powerful computing resources across a new infrastructure called the Grid [1]. Scientists will then have access to very large data sets and be able to perform real-time experiments on this data. This "new" science, which is performed through distributed global collaborations, using very large data collections, terascale computing resources, and high-performance visualizations is called e-Science[5]. So, scientists using e-Science need storage for large files, computing resources, high-performance visualization, and automated workflows. These issues are the same ones that have either been solved or will be solved in the Grid[1].

HVEM[2] lets scientists realize 3-dimensional structure analysis of new materials both in atomic and micrometer-scale utilizing its atomic resolution high

W. Dubitzky et al. (Eds.): GCCB 2006, LNBI 4360, pp. 175–190, 2007.

tilting capability. The type of HVEM referred to in this paper is Transmission Electron Microscope (TEM), which produces images that are projections of the entire object, including its surface and the internal information. When tilting is needed, the gonioMeter rotates to change the angle of the sample. A Charge Coupled Device(CCD) camera produces the actual image inside the HVEM. Users control the HVEM in a fine-grained manner, observe the specimen using the above subcomponents of HVEM, and manipulate high-resolution images using computing/storage resources. To allow remote users to perform the same tasks as on-site scientists, HVEM-Grid provides remote users with applications for controlling the HVEM, observing the specimen, and carrying out other tasks remotely.

The objective of this project is to work towards an implementation of the vision of being able to use the HVEM[1] at Korea Basic Science Institute(KBSI)[2] remotely so that remote users can also carry out large-scale research.

In this paper, we propose and describe an HVEM DataGrid system, focusing on biologic experiments with HVEM. But, because the HVEM DataGrid has a general architecture that is not limited to biological experiments, it can be applied to other fields that require experiments with HVEM, such as materials science and physics. Also, HVEM DataGrid is scalable enough for researchers to share the results of their researches with other scientists, which enables them to extend their research areas.

This paper is organized as follows. Section 2 and 3 briefly review related works and background. We then propose the HVEM DataGird System in Section 4. The implementation details follow in Section 5, and Section 6 concludes the paper with future works.

2   Related Works

Researches using HVEM have been mainly conducted in many fields of natural science, including biology, materials science, and physics as well as Electron Microscopy. In most cases, researchers get their experimental data from an electron microscope (EM) offline and store them in their local storage. When necessary, they must manually move the data to another storage or organize them with other data using software or some other tools. This makes it difficult for researchers to use others' data banks in order to do joint researches or to get help for their researches from the banks.

Many e-Science projects are on-going or being set up. Examples include the cyber-infrastructure implementation in the U.S.[6], the NEESgrid project[10], the e-science project in the U.K.[5][11], and the NAREGI project in Japan[12]. They are large projects supported by their nations with "implementation of an advanced research environment" as the target. NEESgrid has a similar objective as the HVEM Grid system, which is to control special equipment and to experiment using it remotely. NEESgrid has been set up to link earthquake researchers

---

[1] The HVEM in KBSI is JEM/ARM1300s.

[2] www.kbsi.re.kr

across the U.S. with leading-edge computing resources and research equipment, allowing collaborative teams (including remote participants) to plan, perform, and publish their experiments. Besides global services such as resource monitoring and discovery, NEESgrid also provides a data repository to store and search the experimental data as in HVEM Grid. However, NEESgrid is different from HVEM Grid in its subject, equipment for experiments, detailed structure, and subcomponents of the systems, which are better suited to HVEM Grid or NEESgrid according to the object of the systems.

Cyber Databank and DataGrid have been implemented mainly in the field of biology. However, due to the limitations and characteristics in biological experiments, there are many steps that can not be fully automated. NERC Datagrid[31] was constructed with the aim of sharing data for earth systems. CERN DataGrid[25] was implemented to provide intensive computation and analysis of shared large-scale databases, from hundreds of terabytes to petabytes, across widely distributed scientific communities. The former handles atmospheric data and oceanographic data, and the project started in 2002 in England. The latter is for High Energy Physics (HEP) experiments. HEP will start to produce several petabytes of data per year over a lifetime of 15 to 20 years. Since such amount of data was never produced before, special efforts concerning data management and storage are required. There is also the Particle Physics Data Grid(PPDG) project[29], which focuses on developing a Grid infrastructure that can support high-speed data transfers and transparent access.

3   Background

To use the HVEM of KBSI, users must reserve a time slot and visit the laboratory at KBSI, where users can perform the experiment with the help of operators. After the experiment, users can take the results, such as images and video streams, and manipulate the result with the help of programs. Because experimental results and data are mostly obtained offline, they are generally managed by the experimenters privately or by the organization that they belong to. Since the high resolution image files from HVEM are large, it is very burdensome and troublesome to transfer them from a local site to remote sites. As researches are active, valuable data get created. But, the data may have nonuniform types - some data may be large images, others may be large text files, and others may be just information for other data. Therefore, it may be impossible or ineffective for all data to be stored in a single repository, even though it may be scalable with multiple homogeneous storages. In this case, several data repositories are provided for experimental data, users must know all interfaces to the heterogeneous repositories, which may be different. For example, to access a database, users must know the interface to a database. To retrieve data in the file system, they must know its programming interface or need a tool to access it.

So, it is difficult for remote researchers, including fellow researchers, to share the data in near-real-time or to use them as soon as they are produced. This brought about difficulty in group research and expansion of research.

This situation presents two main challenges. One is for users at remote locations to be able to use the HVEM at any time with minimal support from operators. The other is to be able to use the previous experimental results as soon as they are produced, which is the functionality that the HVEM DataGrid system should satisfy. If it is possible to share the experimental results and the related data quickly, other researchers may provide meaningful feedback while the experimenter can still modify the experiment. Also, the research results may be enriched with fast experiments and fast manipulation of data. This requires that our system have the ability to perform remote experiments on the Internet, to manipulate and store results, to search previous experiments, and to show related experimental data to the authorized users. Also, due to a scalable, open architecture, HVEM DataGrid should be applicable or helpful to other scientific researches.

4   HVEM DataGrid System

The overall HVEM Grid[3] system, including HVEM DataGrid, are described in Figure 1. The figure shows its 3-part architecture, which enables it to control the HVEM and CCD camera, to take pictures, to transfer them in real time to the DataGrid, and to manipulate data with necessary software. Therefore, the HVEM-Grid is an integrated system of HVEM - legacy applications for accessing or controlling the HVEM - and a computing/storage resource for HVEM users.

And, this work package will develop a general purpose information sharing solution with unprecedented automation, ease of use, scalability, uniformity, transparency, and heterogeneity.

4.1   System Architecture and Design

Figure 2 describes the abstract functional stack of the HVEM DataGrid. Figure 3 shows the Data Management Server, which is a core part of the HVEM DataGrid. The application area in Figure 2 includes many applications which retrieve, store, or manipulate data. The current implementation of HVEM DataGrid has a simple web interface for storing, retrieving, and searching data. Through authentication, if users are certified, they can use the HVEM DataGrid. Users will be also able to use tomography or other data processing tools through the web interface soon[3]. At that time, through HVEM Grid web portal, users can access and manage data and process it at one site. The Data Grid Services layer manages data by storing it to or retrieving it from a physical storage by logical interfaces. Users can search for data without considering their physical locations or various interfaces to each data storage. The logical interface hides its specific interfaces to physical storage. Currently, we connect Data Grid Services to physical storage using Storage Resource Broker (SRB)[8] except the database for

---

[3] The next version of the HVEM DataGrid will be connected to the tomography so that users can use it online.

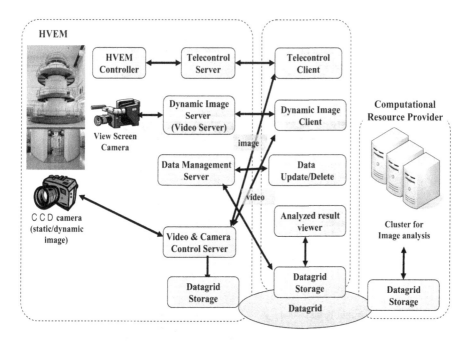

Fig. 1. HVEM Grid System - Overall Architecture

biological information. Because our system is designed to use only the interfaces to read, write, and delete among the SRB interfaces, if other storage management systems can support similar interfaces and the functionality of transparent access to heterogeneous storage, HVEM DataGrid can adopt it in place of SRB.

The storage in HVEM DataGrid consists of a database and a file storage as described in Figure 3. If the experiment using HVEM is related to biology, the relatively small data are collected in the database. Such data include information for materials, material handling methods, experiments, and experimenters. Large data, such as images, documents, and animations, are maintained in the file storage. The Database Management System (DBMS) provides easy interfaces to manage data in the database and ensures relatively fast data search. SRB supports transparent access to heterogeneous storage as well as providing a scalable repository. All data including image files and documents in HVEM DataGrid are searched for in its database. After finding its logical path, i.e., the identification, in the file storage by first searching the database, users can retrieve the files from the file storage, SRB. Although the physical storage locations for the file and the database may be distributed on several machines, the client interface to HVEM DataGrid is unique. So users, including researchers, can consider the several repositories as one big repository. However, access to data in DataGrid is limited according to users' authorization levels. In Figure 3, the data management server has two storage interfaces - the interface to the database, i.e., the JDBC driver, and the interface to file storage, i.e., the SRB

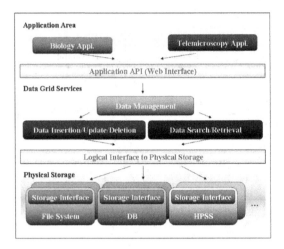

Fig. 2. HVEM DataGrid System

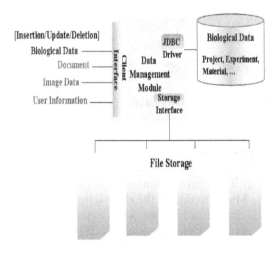

Fig. 3. Data Management Server

interface. The data management server processes client requests through them. Current implementation of our system uses Postgres[7] as its DBMS.

4.2 Schema of HVEM DataGrid

Figure 4 shows the schema of the HVEM DataGrid system. The current version of HVEM DataGrid is specialized for biological experiments using HVEM. A datagrid for other fields using HVEM or TEM can be additionally implemented to the current HVEM DataGrid system, or it can be newly constructed like the HVEM DataGrid system.

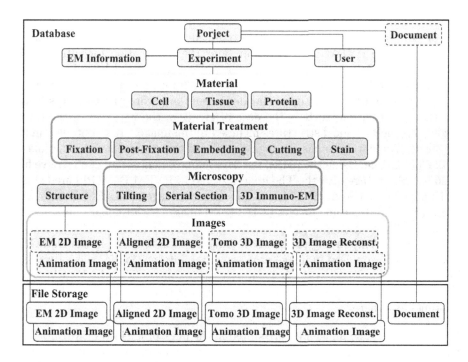

Fig. 4. Schema of HVEM DataGrid

Because the HVEM DataGrid maintains biological information, its data schema should be organized to suitably express biological meaning. We ensure that HVEM DataGrid accommodates these requirements precisely. As shown in Figure 4, the upper box describes the schema on the relational database and the lower box expresses the file storage.

Relational Database. All biological information is maintained and managed hierarchically in the database of the HVEM DataGrid. Projects exist at the top level of the hierarchy, and several experiments constitute a project. Experiments are created according to the materials used in experiements with HVEM. The information for an experiment is stored in the DataGrid with the information of HVEM, which is used for the experiment. When a user searches for the information of an experiment, all information related to the experiment, such as the HVEM information, material, and experimenter, is retrieved together. All documents are stored with its related projects. So when users search for a project, they can read the information for related documents and download them or peruse the information on the project. Three categories of the materials maintained in the first version of HVEM DataGrid are cell, tissue, and protein. Users must choose one category their material belongs to. The methods for material handling are fixation, post-fixation, embedding, cutting, and stain. All or some of these methods can be applied to the material. According to microscopy, the HVEM DataGrid maintains the information for tilting, serial section, and 3-D Immuno-EM. The information for material handling and microscopy can be retrieved related with the search for an experiment or the material. Even though image data and documents are stored at the file storage, the their information, such as the date of creation, the file creator, the tilting angle, and the biological structures included in the image, are maintained in the database. The database has the logical paths of the documents and the images in the file storage. Using the paths, the data management server can retrieve files from the file storage directly. The images stored in HVEM DataGrid are (1) 2-D EM images taken from HVEM, (2) 2-D aligned images (2-D EM images with alignment), (3) 3-D tomo images (2-D aligned images with tomography), and (4) 3-D reconstruction images (3-D tomo images with polygonal modeling). Animation images corresponding to the above images are also stored in the DataGrid. If there is software used in image processing, the information for the softwares is also added to the DataGrid. So when users have images from the HVEM Data-Grid and the software used to manipulate the images, they can reproduce them anytime, anywhere.

File Storage. The image files are stored in the file storage in HVEM DataGrid, where they are classified according to each experiment. Documents are stored according to its corresponding project. Because the file storage in HVEM Data-Grid is scalable with the support of SRB, physical storage can be distributed across the network. The HVEM DataGrid can also work with other storage management systems because it does not depend on any specific functionality provided by SRB.

4.3   Data Workflow

Figure 5 shows the data workflow in the HVEM DataGrid system. As a unified data management system, HVEM DataGrid should do a systematic management for all data stored in it. For example, EM images from HVEM don't have any meaning without the biological information related to the images, such as for material, microscopy, experimental information. After we store all biological information for the images, we should keep the images in HVEM DataGrid. Therefore, it is very important to represent the relations and the storage precedence among data in HVEM DataGrid. Because data in HVEM DataGrid are not managed in a homogeneous storage such as a database, the relational semantics between various data stored in heterogeneous storages should be supported or traced by the HVEM DataGrid. HVEM DataGrid does that through the data workflow. According to the data workflow, users insert, update, and delete data in the HVEM DataGrid. It is needless for users to check or keep in mind where they stop data insertion. The next time they access the HVEM DataGrid, they can know where the data insertion stopped and proceed from there. The overall data workflow process is as follows:

User interfaces for the insertion and modification of data to the database and the file storage are unified. When a user logs in and is about to create a project or an experiment, he can see all the information for the projects or the experiments that he has previously created or modified. Then, he can choose any project or experiment listed to modify it or to fill the information omitted in its table. He can create a new project or experiment other than the ones listed. The user must choose to either modify or create at every step until the completion of a database instance for a project or experiment. Users can also continue previous insertion of data; they can proceed data insertion or modification from where they left off. The data workflow described in Figure 5 expresses data modification as backward arrows and stopped data insertion as discrete block arrows. However, the order in which data is inserted into the DataGrid, which is expressed in Figure 5 as the direction of arrows, should be maintained. Because the order is kept by HVEM DataGrid, users need not be concerned.

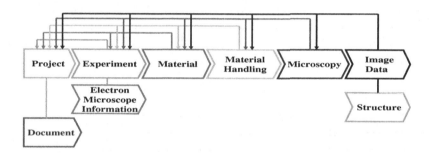

Fig. 5. Data Workflow in HVEM DataGrid

After a user creates a project on the DataGrid, he can create an experiment or store documents related to the project. Several experiments can be created under one project. When a user inserts the information for an experiment, he should fill the data for the electron microscope. He can stop data insertion and log out at any time because the data he inserts are stored in the DataGrid. The user must choose one category among cell, tissue, and protein as the material used in the experiment. Next, the user can insert the information for material handling methods (fixation, post-fixation, embedding, cutting, stain) before taking a picture with HVEM. Next, the information for the microscopy related to the experiment (tilting, serial section, 3-D Immuno-EM) should be stored. In particular, 3D Immuno-EM is related to the handling of transfection and antibody, so these information should be inserted with 3D Immuno-EM. The user can then upload image files to the storage by inserting their information to the database. EM 2-D images are usually composed of tens or hundreds of high-resolution images with sizes between 2MB and 3MB. To upload these images to the storage at once, HVEM DataGrid supports directory uploads as well as file uploads.

4.4   Data Management and Access

Users have different authorizations for data access according to their level. If a user is categorized as a manager, he can create a project and insert all data related to it to the DataGrid. If a user is classified as researcher, he can not create a project, but he can create a new experiment under an existing project. The operator can only insert image information and upload them to the file storage. Data deletion can be executed only by the creator and the project manager. Other users can only read the data maintained in the DataGrid. The reason for limiting users' authorization for data in the DataGrid is data protection. Because it is relatively easy for any user to access the HVEM DataGrid through the Web interface, if everyone can modify or create data, there is a chance that the HVEM DataGrid will be damaged. Therefore, HVEM DataGrid allows only a few to have high-level authorizations. Currently, HVEM DataGrid categorizes users into only three groups: manager, researcher, and operator. But in the next version, we will refine the user groups based on real use cases from the field.

4.5   Data Search

The SRB browser is an application to show files or directories in the file storage much like the Microsoft Windows explorer. A browser for the file storage in HVEM DataGrid will be developed next year. Up until now, researchers have stored and manipulated data offline. But, independent image files are meaningless if they are not related to other experimental information. So, the browser may be helpful to offline work if the researchers have or know the related information well. The researchers may use the browser for the file storage in a similar manner as the SRB browser in order to put data to or to get data from the file storage directly, not through the DataGrid which integrates a database and a file storage. But in this case, it is difficult for the researchers to carry out integrated management for the image files and the related data. So the browser in HVEM DataGrid will be designed to grant only a limited rights to users.

The search mechanism in HVEM DataGrid was designed to find all information for experiments or projects in its database. Because the information for the files stored in the storage can be found in the database, the search rate for some data is dependent on the DBMS that HVEM DataGrid uses. The data search service is provided by two interfaces. One is a random search and the other is a systematic one. The former searches for data using a keyword in one category and shows all data related to the result. The categories users can choose from are project, experiment, material, document, and user. Random searches not confined to any search category is not allowed because of the long delay that results from searching all data tables in the database. The latter provides users a fast search due to a more refined search condition.

5   System Implementation

In this section, we will describe the implementation of the HVEM DataGrid system and its test results. The current implementation of HVEM DataGrid is the first version, and it will be expanded with a browser for file storage and a connection to a computational grid next year.

5.1   Environmental Setup

HVEM DataGrid consists of a web server, a database server, and a file storage. The web server and the database server are Pentium IV 3.0GHz machines with 1GB of RAM, and a 250GB HDD. The file storage consists of two Pentium IV 3.00GHz machines with 1G of RAM and a 250GB HDD. All machines use Redhat 9 as their OS. Because the file storage is scalable in size, if necessary, other machines can be added in order to extend the total size of file storage without any negative effects.

5.2   DataGrid Test

We tested HVEM DataGrid for data creation, modification, deletion, and search. Figure 6 shows the main screen of the HVEM DataGrid. The left picture shows the pages before logging in and the right one shows the page after logging in.

Data Creation/Update. Figure 7 describes the interface of data insertion and modification for "project." The figure shows that next to project, users should

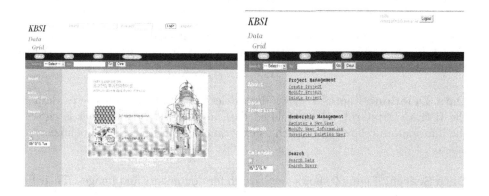

Fig. 6. The main screen of HVEM DataGrid

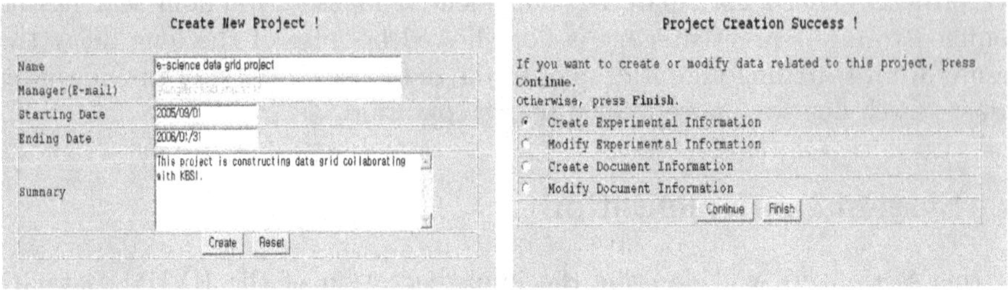

(a) Data Creation

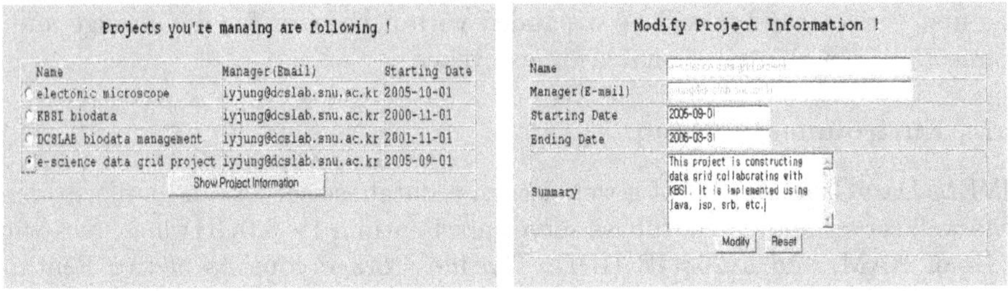

(b) Data Modification

Fig. 7. Data Creation/Modification

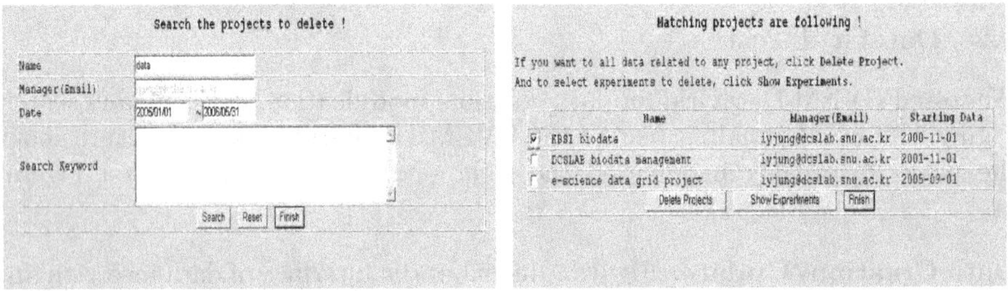

(a) Search condition for data deletion   (b) Search results for data deletion

Fig. 8. Data Deletion

manage data on experiments or documents as discussed in the section on Data Workflow. With the same interface, users can manipulate other data in HVEM DataGrid.

Data Deletion. Figure 8 shows data deletion. When a user deletes data in the HVEM DataGrid, all the related data are deleted. If a project is chosen, the information related to it, such as experiments, materials, and documents, is deleted.

Data Search. Figure 9 shows data search for documents and images. Their full or partial information can be displayed according to users' requests. They can also be downloaded.

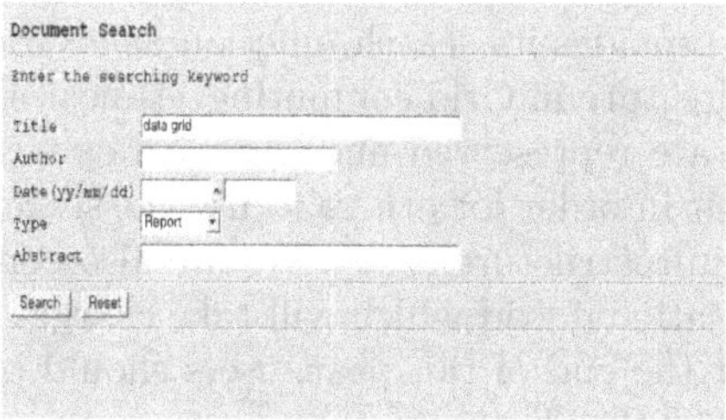

(a) Search condition for documents

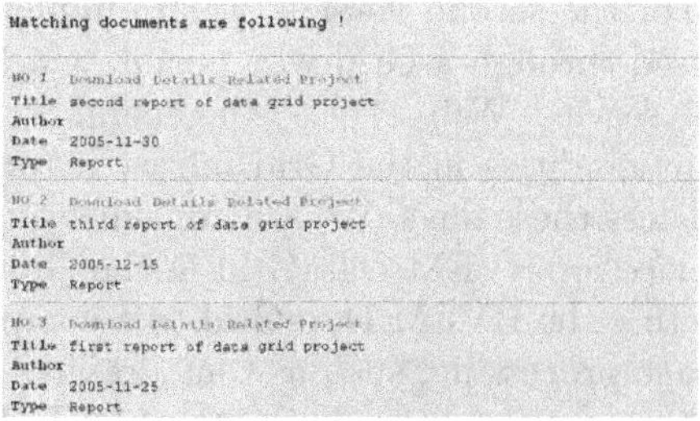

(b) Search results for Documents

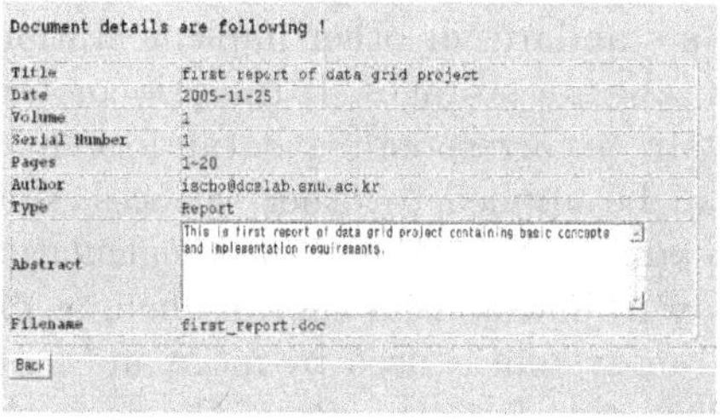

(c) Search results for one document in detail

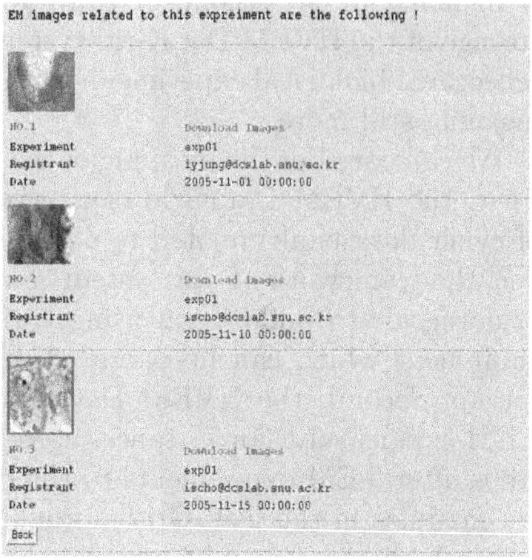

(d) Search results for images in detail

Fig. 9. Data Search

6   Conclusion and Future Work

In this paper we explained HVEM DataGrid, designed especially for biological experiments, which provides many capabilities required by remote researchers. The primary goals of this system are to enable individual scientists to perform large-scale research and to advance the formation of various research communities through the proposed infrastructure. To do so, remote users should be able to access both the large images and their related information easily. We believe our system can satisfy the capabilities listed above.

The first version of HVEM DataGrid has been implemented, and we are currently testing it with real image data from HVEM experiments and its related biological data. The image data files from HVEM are so large that if users upload them to the file storage in the HVEM DataGrid at once, they may encounter a long upload delay. We are studying the performance of file transfers so that the upload or download delay of large files become tolerable. We are also considering other functions that can be added for users' convenience.

For the second version, we are implementing the HVEM DataGrid to be accessed through a Grid web portal, which provides its search function as a Grid web service. Web service is a general architecture in Grid computing. Almost all functionalities in the Grid infrastructure are represented and provided as web services these days. We are implementing it in order for others to use the search service for HVEM DataGrid in the Grid infrastructure easily. We are also connecting the HVEM DataGrid to a computational grid which will take charge of image processing, such as tomography. At the end of this year, users should be able to use the second version of HVEM DataGrid.

Included in the second or third version will be a browser to access the file storage of the HVEM DataGrid system. We will compensate the data schema to reflect real biological experiments with HVEM and to precisely express biological meaning still more.

We can consider the meaning of the HVEM DataGrid from several viewpoints. First, the HVEM DataGrid can serve as motivation to other implementations of cyber data banks related to natural science or engineering. As described earlier, the framework of our system can be a standard for other implementations of datagrids to build upon using various types of storages. If there is another equipment which can be operated remotely, a corresponding datagrid can be set up. Second, the HVEM DataGrid can be applied to other sciences using TEM expansively. For instance, users can add data manipulation or calculation steps after TEM experiments to the data workflow and implement them as extra modules in the DataGrid system. As our system is used by more and more researchers, there may be additional requirements for our system. Owing to the additional implementation and compensation, our system will be stable, and we are sure that our system will allow more and more scientists to achieve successful researches.

References

1. Foster, I. and Kesselman, C. and Tuecke, S., The anatomy of the grid: Enabling scalable virtual organizations, *Intl. J. Supercomputer Applications* (2001)
2. Korea Basic Science Institute, KBSI Microscopes & facilitates, *http://hvem.kbsi.re.kr/eng/index.htm*
3. Hyuck Han, Hyungsoo Jung, Heon Y. Yeom, Hee S. Kweon and Jysoo Lee, HVEM Grid: Experiences in constructing an Electron Microscopy Grid *The Eighth Asia Pacific Web Conference* (2006)
4. Globus Aliiance, Globus Toolkit 4.0 Release Mannuals, *http://www.globus.org/toolkit/docs/4.0*
5. Oxford e-Science Centre, e-Science Definitions, *http://e-science.ox.ac.uk/public/general/definitions.xml*
6. Daniel E. Atkins et al, Revolutionizing Science and Engineering Through Cyber-infrastructure: Report of the National Science Foundation Blue-Ribbon Advisory Panel on Cyberinfrastructure, *National Science Foundation*(2003)

7. PostgreSQL Global Ddevelopment Grpoup, PostgreSQL 8.0,
   http://www.postgresql.org/
8. San Diego Supercomputer Center, SDSC Storage Resource Borker(SRB),
   http://www.sdsc.edu/srb
9. Martone ME, Zhang S, Gupta A, Qian X, He H, Price DL, Wong M, Santini S,
   Ellisman MH, The cell-centered database: a database for multiscale structural and
   protein localization data from light and electron microscopy *Neuroinformatics, 1(4)*
   (2003) 235-245
10. NEESgrid, *http://www.neesgrid.org*
11. UK e-science Programme, *http://www.nesc.ac.uk*
12. Japan NAREGI project, *http://www.naregi.org*
13. Korean Society of Electron Microscopy, *http://ksem.com/*
14. European Microscopy Society, *http://www.eurmicsoc.org/*
15. Microscopy Society of America, *http://www.microscopy.org/*
16. The Royal Microscopical Society, *http://www.rms.org.uk/*
17. Telescience Project *https://telescience.ucsd.edu*
18. M. Hadida-Hassan, S.J. Young, S.T. Peltier, M. Wong, S.P. Lamont, M.H. Ellisman,
    Web-based telemicroscopy *Journal of Structural Biology, 125(2/3)* (1999) 235-245
19. Peltier, S.T., et al, The Telescience Portal for Advanced Tomography Applications
    Journal of Parallel and Distributed Computing: Computational Grid, 63(5) (2002)
    539–550
20. Lin, A.W., et al, The Telescience Project : Applications to Middleware Inter-
    action Components *Proceedings of The 18th IEEE International Symposium on
    Computer-Based Medical Systems* (2005) 543–548
21. Biogrid Project *http://www.biogrid.jp/*
22. Chaos Project *http://www.cs.umd.edu/projects/hpsl/chaos/*
23. Asmara Afework, Michael Beynon, Fabian Bustamante, Angelo Demarzo, Renato
    Ferreira, Robert Miller, Mark Silberman, Joel Saltz, Alan Sussman and Hubert
    Tsang, Digital Dynamic Telepathology - the Virtual Microscope *Proceedings of the
    1998 AMIA Annual Fall Symposium, American Medical Informatics Association,
    Hanley and Belfus, Inc.* (1998)
24. Renato Ferreira, Bongki Moon, Jim Humphries, Alan Sussman, Joel Saltz, Robert
    Miller and Angelo Demarzo, The Virtual Microscope *Proceedings of the 1997 AMIA
    Annual Fall Symposium, American Medical Informatics Association, Hanley and
    Belfus, Inc.* (1997)
25. CERN Datagrid Project *http://eu-datagrid.web.cern.ch/eu-datagrid/*
26. V. Breton, R. Medina, J. Montagnat, DataGrid, Prototype of a Biomedical Grid,
    Methods of Information in Medicine, 42(2), Schattauer(2003)143–148
27. Wolfgang Hoschek, Javier Jaen-Martinez, Asad Samar, Heinz Stockinger, Kurt
    Stockinger, Data Management in an International Data Grid Project, *IEEE/ACM
    International Workshop on Grid Computing Grid'2000, Bangalore, India* (2000)
28. H.Stockinger, F.Dono,E.Laure, S.Muzzafar et al, Grid Data Management in action,
    computing in high energy physics (CHEP 2003)
29. Paricle Physics Data Grid *http://www.ppdg.net/*
30. Kurt and Heinz Stockinger, L.Dutka, R.Slota et al, Acces Cost Estimation for
    Unified Grid Storage Systems, *IEEE/ACM 4th international workshop on grid
    computing (Grid2003), IEEE Computer Society Press*

31. NERC DataGrid Web Site *http://ndg.badc.rl.ac.uk/*
32. Lawrence B.N., R. Cramer, M. Gutierrez, K. Kleese van Dam, S. Kondapalli, S. Latham, R. Lowry, K. O'Neill and A.Woolf, The NERC DataGrid Prototype, *Proceedings of the U.K. e-science All Hands Meeting, S.J.Cox(Ed) ISBN 1-904425-11-9* (2003)
33. Stephens A., Marsh K.P. and B.N. Lawrence, Presenting a multi-terabyte dataset via the web, *Proceedings of the ECMWF Ninth Workshop on Meteorological Operational Systems* (2003)

Author Index

Lecture Notes in Computer Science

For information about Vols. 1–4257

please contact your bookseller or Springer